Cutting Edge Nanotechnology

Cutting Edge Nanotechnology

Editor

Rodolph Donovan

scitus
academics

Cutting Edge Nanotechnology
Edited by **Rodolph Donovan**

ISBN: 978-1-68117-315-3
Library of Congress Control Number: 2016937625

SCITUS Academics LLC,
www.scitusacademics.com
Box No. 4766, 616 Corporate Way,
Suite 2, Valley Cottage,
NY 10989

Notice

Printed in United States of America on Acid Free Paper

Preface

Nanotechnology is the study and application of very small subjects and can be used in all other science areas, such as chemistry, biology, physics, materials science and technology. In its original sense, nanotechnology refers to the projected ability to build items from below using the techniques and tools developed today to produce complete, high-performance products. This world is the atomic and nanostructural world of nanotechnology. Nanotechnology is so new; nobody knows what will come out of it. However, the predictions differ from the ability of nanobots to reproduce things like diamonds and food in the world. Scientists are currently talking about the future effects of nanotechnology. Nanotechnology can create many new materials and equipment for a wide range of applications, e.g. in medicine, electronics, biomaterial energy production and consumer products. Nanotechnology, on the other hand, raises many of the same issues as any new technology, including concerns about nanomaterials' toxicity and environmental impacts and their potential impact on the global economy.

The primary goal of this book is to describe concerns that are important in innumerable types of devices—from conventional transistors to molecular electronic devices. Current work and more advanced concepts are covered. The book will serve as a guide for identifications of very important areas of investigation in micro-, molecular electronics and nanotechnology.

Table of Contents

Chapter 1

Study of Materials Deformation in Nanometric Cutting by Large-scale Molecular Dynamics Simulations

QX Pei, C Lu, HP Lee and YW Zhang

[1]*Institute of High Performance Computing*

ABSTRACT

Nanometric cutting involves materials removal and deformation evolution in the surface at nanometer scale. At this length scale, atomistic simulation is a very useful tool to study the cutting process. In this study, large-scale molecular dynamics (MD) simulations with the model size up to 10 millions atoms have been performed to study three-dimensional nanometric cutting of copper. The EAM potential and Morse potential are used, respectively, to compute the interaction between workpiece atoms and the interactions between workpiece atoms and tool atoms. The material behavior, surface and subsurface deformation, dislocation movement, and cutting forces during the cutting processes are studied. We show that the MD simulation model of nanometric cutting has to be large enough to eliminate the boundary effect. Moreover, the cutting speed and the cutting depth have to be considered in determining a suitable model size for the MD simulations. We have observed that the nanometric cutting process is accompanied with complex material deformation, dislocation formation, and movement. We find that as the cutting depth decreases, the tangential cutting force decreases faster than the normal cutting force. The simulation results reveal that as the cutting depth decreases, the specific cutting force increases, i.e., “size effect” exists in nanometric cutting.

Keywords: Molecular dynamics Nanometric cutting Materials deformation Large-scale simulation

INTRODUCTION

Nanometric cutting is a tool-based materials removal technique to remove materials at nanometer scale thickness in the surface. Nanometric cutting can be used to produce micro/nano-components with nanoscale surface finish and sub-micron level form accuracy for many applications such as micro-electro-mechanical systems (MEMS) and nano-electro-mechanical systems (NEMS) [1, 2]. Understanding the material removal mechanism and mechanics at atomistic scale in the surface, such as deformation evolution, chip formation, machined surface, cutting forces, and friction, is a critical issue in producing high precision components. However, as the nanometric cutting process involves only a few atomic layers at the surface, it is extremely difficult to observe the cutting process and to measure the process parameters through experiments. Therefore, theoretical analysis plays a major role in obtaining information on nanometric cutting. The widely used finite element method based on continuum mechanics for the analysis of conventional cutting is not appropriate to analyze the nanometric cutting process because of the discrete nature of materials at such a small length scale; therefore molecular dynamics (MD) simulation has become a very useful tool in the study of nanometric cutting.

A number of studies have used the MD simulations to analyze the nanometric cutting process [3–9]. The typical studies among them include: Maekawa et al. [3] studied the role of friction between a single-crystal copper and a diamond-like tool in nano-scale orthogonal machining. The Morse type potentials were used for the interactions between Cu–Cu, Cu–C, and C–C atoms; Zhang et al. [4] studied the wear and friction on the atomic scale and identified four distinct regimes of deformation consisting of no-wear, adherence, plowing, and cutting regimes; Komanduri et al. [5–7] carried out MD simulations of nanometric cutting of single-crystal copper and aluminum. They investigated the effects of crystal orientation, cutting direction and tool geometry on the nature of deformation, and machining anisotropy of the material; more recently, Zhang et al. [9] used MD simulations to study the subsurface deformed layers in the atomic force microscopy (AFM)-based nanometric cutting process.

All those previous studies have provided much help in understanding nanometric cutting. However, as the MD simulation of nanometric cutting is compute-intensive, small simulation models with a few thousands to tens of thousands of atoms were used in the reported studies to reduce the computing time. Although those small models have provided a lot of information on the nanometric cutting processes, a small model may induce significant boundary effects that make the results unreliable. For example, if the model is not large enough, the widely used fixed-atoms boundary in MD simulations may have strong effect on the dislocation movement and thus will affect the motion of atoms at the cutting surface. Besides, in most of the reported studies, the simulation models are two-dimensional or quasi-three-dimensional (plane strain) due to the limitation on the model size. Therefore, there is a need for large-scale MD simulations of three-dimensional (3D) nanometric cutting processes.

Another limitation of previous studies on MD simulations of nanometric cutting of metals is that the Morse potential has been widely adopted to model the interatomic force between metal atoms. Morse potential is a pair potential which considers only two-body interactions; thus, it provides a rather poor description of the metallic bonding. The strength of the individual bond in metals has a strong dependence on the local environment. It decreases as the local environment becomes too crowded due to the Pauli's "exclusion principle" and increases near surfaces and in small clusters due to the localization of the electron density. The pair potential does not depend on the environment and, as a result, cannot reproduce some of the characteristic properties of metals, such as the much stronger bonding of atoms near surfaces. The EAM potential, which has been specially developed for metals [10–12], can better describe the metallic bonding. Therefore, the EAM potential gives a more realistic description of the behavior and properties of metals than the Morse potential. Our previous study [13] showed that the two different potentials resulted in quite different simulation results and suggested that the EAM potential should be used in MD simulation of nanometric cutting.

In this article, we present large-scale 3D MD simulations of nanometric cutting of copper. In our simulations, the EAM potential is employed for the interactions between Cu atoms in the workpiece. We first studied the model size effect on the simulation results with three different model sizes of about 2, 4, and 10 million atoms. Then, we used the 4-million-atom model, which is shown to be large enough to eliminate the boundary effect, to study the detailed materials deformation, dislocation movement, and cutting forces during the cutting processes.

Simulation Models and Conditions

Figure 1a–c show three simulation models for our large-scale MD simulations of nanometric cutting. The workpiece sizes are 40 × 20 × 30 nm containing 2,053,594 atoms, 40 × 40 × 30 nm containing 4,098,686 atoms, and 70 × 44 × 40 nm containing 10,137,600 atoms. The diamond tool contains 8446 carbon atoms. The cutting is along the*x*direction, which is taken as the [100] direction of the FCC lattice of copper. The boundary conditions of the cutting simulations include: (1) three layers of atoms at the bottom of the workpiece materials (lower*z* plane) are kept fixed; (2) periodic boundary conditions are maintained along the*y* direction.

In nanometric cutting, as the cutting depth can be as small as a few nanometers, the edge of the cutting tool is not sharp compared with this very small cutting depth. The edge radius of the cutting tool is usually much larger than the cutting depth. Therefore, in our large-scale MD simulations, we use a round edge cutting tool with an edge radius of 6 nm instead of a sharp cutting tool. The geometry of the cutting tool is shown in Fig. 2. The tool thickness is 3.2 nm with the tool rake angle α and the tool clearance angle β being 12°.

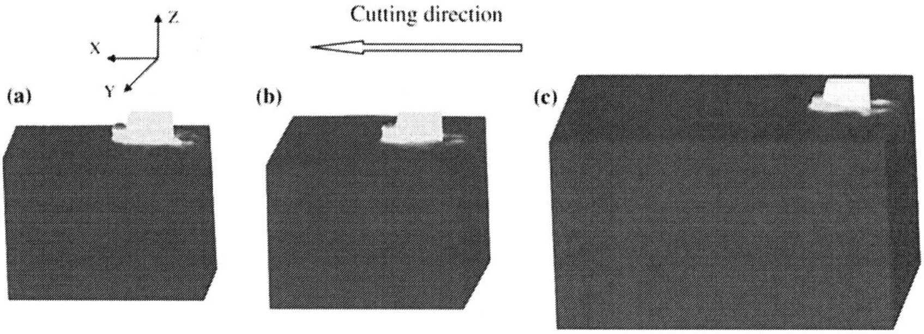

Figure 1. The MD simulation models with the number of atoms in the workpiece being around**a** 2 millions,**b** 4 millions and**c** 10 millions. The corresponding workpiece dimensions are 40 × 20 × 30 nm, 40 × 40 × 30 nm, and 70 × 44 × 40 nm, respectively. The cutting tools are in light grey color and the cutting chips ahead the cutting tools are shown in colors ranging from red to light blue

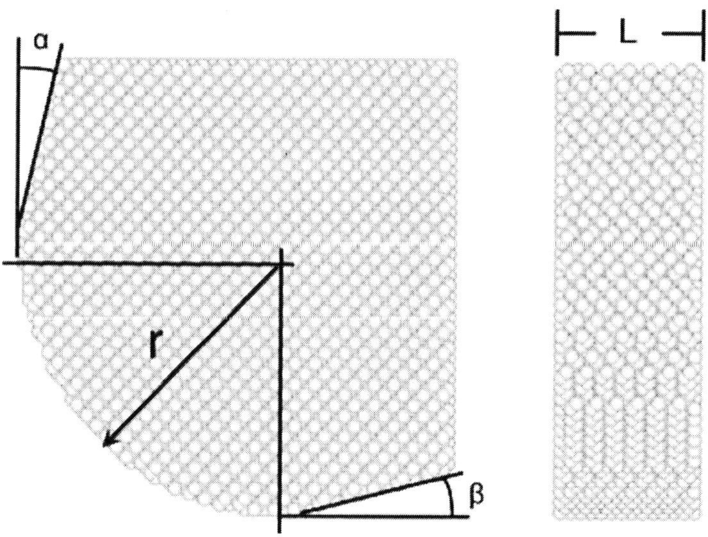

Figure 2. The geometry of the cutting tool. The tool edge radius$r$ = 6 nm. The rake angle α = 12° and clearance angle β = 12°. The tool thickness$L$ = 3.2 nm

The cutting speed used in the MD simulations ranges from 50 to 500 m/s, while the cutting depth ranges from 0.8 to 4 nm. The cutting is in the (001) plane and along the [100] direction of the workpiece. The initial temperature of the workpiece is 300 K. The three layers of atoms adjacent to the fixed-atom boundary at the workpiece bottom are set as the thermostat atoms, in which the temperatures are maintained at 300 K by rescaling the velocities of the atoms. The velocity Verlet algorithm with a time step of 2 fs is used for the time integration of Newton's equations of motion.

The interatomic forces in MD simulations are calculated from the interatomic potentials. The Morse potential is relatively simple and computationally inexpensive compared to the EAM potential. The Morse potential is as follows:

$$\varphi(r_{ij}) = D\left\{\exp\left[-2\alpha\left(r_{ij} - r_0\right)\right] - 2\exp\left[-\alpha\left(r_{ij} - r_0\right)\right]\right\} \tag{1}$$

where $\varphi(r_{ij})$ is a pair potential energy function; D is the cohesion energy; α is the elastic modulus; r_{ij} and r_0 are the instantaneous and equilibrium distance between atoms, i and j, respectively.

The EAM method, which has been evolved from the density-function theory, is based upon the recognition that the cohesive energy of a metal is governed not only by the pair-wise potential of the nearest neighbor atoms, but also by embedding energy related to the "electron sea" in which the atoms are embedded. For EAM potential, the total atomic potential energy of a system is expressed by the following equation:

$$E_{tot} = \frac{1}{2}\sum_{i,j} \Phi(r_{ij}) + \sum_{i} F_i(\overline{\rho}_i) \tag{2}$$

where $\Phi_{ij}(r_{ij})$ is the two-body interaction energy between atoms,*i* and*j*, with separation distance,r_{ij};F_i is the embedding energy of atom,*i*; $\bar{\rho}_i$ is the host electron density at site,*i*, induced by all other atoms in the system, which is given by the following equation:

$$\bar{\rho}_i = \sum_{j \neq i} \rho_j(r_{ij}) \quad (3)$$

where $\rho_j(r_{ij})$ is the contribution to the electronic density at atom,*i*, due to atom,*j*, at distance,r_{ij}, from the atom,*i*.

There are three different atomic interactions in the MD simulations of nanometric cutting processes: (1) the interaction in the workpiece; (2) the interaction between the workpiece and the tool; and (3) the interaction in the tool. For the interaction between the copper atoms in the workpiece, we used the EAM potential for copper constructed by Johnson [14]. For the interaction between the copper workpiece and the diamond tool, as there is no available EAM potential between Cu and C atoms, we still use the Morse potential for the workpiece–tool interaction with the parameters adopted from reference [4] being D = 0.087 eV, α = 5.14, and r_0 = 2.05 Å. Since the diamond tool is much harder than the copper workpiece, it is a good approximation to take the tool as a rigid body. Therefore, the atoms in the tools are fixed relative to each other, and no potential is needed for the interaction among the tool atoms.

Dislocations play a crucial role in the plastic deformation of materials. However, accurately identifying dislocations at room temperature in MD simulations is a very difficult task due to thermal vibration of atoms. This might be the reason why almost all the previous MD studies of dislocations were carried out at extremely low temperature of 0 K or 1 K [15–21]. The widely used methods to identify dislocations and other lattice defects in MD simulations are the atomic coordinate number [15], the slip vector [16], and the centro-symmetry parameter [17]. We compared these different methods and found that the methods of atomic coordinate number and the slip vector would become less effective in identifying the lattice defects at finite temperature due to thermal fluctuations of atoms. Therefore, we have chosen to use the centro-symmetry parameter, which is less sensitive to the temperature increase. In a centro-symmetric material (such as copper and other FCC metals), each atom has pairs of equal and opposite bonds among its nearest neighbors. As the material is distorted, these bonds will change direction and/or length, but they will remain equal and opposite under homogeneous elastic deformation. If there is a defect nearby, however, this equal and opposite relation no longer holds. In a perfect bulk FCC lattice, each atom has 12 nearest-neighbor bonds or vectors. The centro-symmetry parameter for each atom is defined as follows:

$$CSP = \sum_{i=1,6} \left| \vec{R}_i + \vec{R}_{i+6} \right|^2 \tag{4}$$

where R_i and R_{i+6} are the vectors corresponding to the six pairs of opposite nearest neighbors in the FCC lattice. By definition, the centro-symmetry parameter is zero for an atom in a perfect FCC material under any homogeneous elastic deformation and non-zero for an atom which is near a defect such as a cavity, a dislocation, or a free surface.

The large-scale MD simulations of nanometric cutting are carried out on the IBM p575 supercomputer at the Institute of High Performance Computing (IHPC). The multi-processor parallel computing is used for the simulations. The parallel computing is realized by using message passing interface (MPI) library. The calculation time for each simulation case depends on the model size, cutting speed, cutting distance, as well as the number of CPUs used. For example, it took about 3 weeks to finish the simulation run for the 10-millino-atom model with the cutting speed of 100 m/s using 32 CPUs.

SIMULATION RESULTS

The Simulation Model Size

For a MD simulation, the larger the model size, the less obvious the boundary effect on the simulation results. However, a very large model will take unnecessarily long computing time. Therefore, it is necessary to study the model size effect, so that we can find a suitable model size for the MD simulations of nanometric cutting. The model size should be moderate with diminished boundary effect on the simulation results.

We first performed MD simulations using the 2-million-atom model in Fig. 1a with a cutting speed of 100 m/s and a cutting depth of 4 nm. The simulation results of the 2-million-atom model are shown in Fig. 3a, from which one can see that the lattice defects generated from the cutting exist in the whole subsurface region between the periodic boundaries (see the front view). The centro-symmetry parameter (CSP) is used to identify the lattice defects. In Fig.3a–c, the atoms inside the model with CSP smaller than three are all eliminated in the visualizations, as these atoms are assumed to be in perfect FCC configuration. Note that the isolated atoms distributed inside the model are not lattice defects. Those atoms having CSP above three are due to the thermal vibration of atoms at finite temperature. The periodic boundary condition in *y* direction implies that both the workpiece and the cutting tool repeat in this direction. The repeated cutting tools may make the stresses at the periodic boundary regions higher due to stress superposition arising from the interaction of stress fields. The stress interaction is helpful for the dislocations in the cutting regions to slide to the periodic boundaries and also helpful for new dislocations to be generated at the periodic boundaries. This phenomenon was also reported by Saraev et al.[21] in their study of the nanoindentation of copper. As lattice defects exist in the periodic boundaries in the simulation results, the 2-million-atom model is not large enough to eliminate the boundary effect at the periodic boundaries, though it is quite large compared with the models used in the reported works on MD simulation of nanometric cutting.

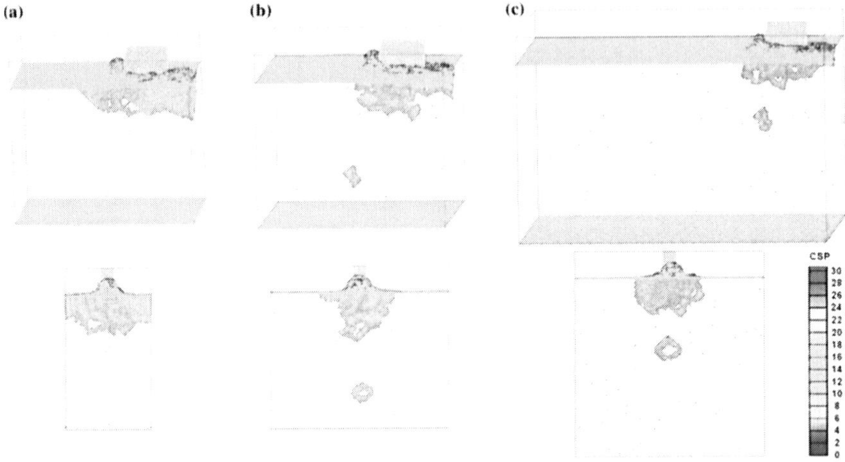

Figure 3. The simulation results of the different model sizes:**a** 2-million-atom model,**b** 4-million-atom model and**c** 10-million-atom model. The lower figures are front views of the models. The cutting depth is 4 nm and the cutting speed is 100 m/s. The blue color shows the dislocations formed inside the workpieces during cutting

Thereafter, we performed simulations using the 4-million-atom model in Fig. 1b with the workpiece thickness (*y*direction) two times that of the 2-million-atom model. The simulation results in Fig. 3b show that the 4-million-atom model could eliminate the boundary effect of the periodic boundaries. We also carried out simulations with the 10-million-atom model in Fig. 1c. In the 10-million-atom model, the workpiece is larger than that of the 4-million-atom model in all the three dimensions with very obvious increase in both the*x* and*z* directions to test the boundary effects in these two directions. We found that the simulation results with the 10-million-atom model, shown in Fig. 3c, did not show obvious difference from those of the 4-million-atom model. Therefore, for the cutting speed of 100 m/s and cutting depth of 4 nm, the 4-million-atom-model is shown to be large enough to ignore the boundary effect in the simulations.

MD simulations were also carried out to study the effect of cutting speed and cutting depth on the boundary effect. The simulation results show that reducing cutting speed results in more obvious boundary effect, while reducing cutting depth results in less obvious boundary effect. This is because a slower cutting speed means longer cutting time, and therefore the dislocations have more time to move and are more possible to reach the boundaries, which makes the boundary effect stronger. A smaller cutting depth means less material deformation, and therefore results in a weaker boundary effect. As the cutting speed and cutting depth may make the boundary effect stronger, it is important to consider those process parameters in choosing the model size for MD simulations of nanometric cutting.

Material Deformation, Dislocations, and Cutting Forces

We now analyze the nanometric cutting process of the 4-million-atom model. Figure 4a–c show the cross-sectional views of the*x*–*z* plane at three different cutting distances of 8, 12, and 16 nm, respectively. In this simulation case, the cutting speed is 100 m/s and the cutting depth is 0.8 nm. It can be seen from the figures colored by CSP that the workpiece materials deform during cutting and the material removal takes place via the chip formation as in conventional cutting. The materials in front of and beneath the tool are away from the perfect FCC lattice. Dislocations and other lattice defects are generated in these regions. It can be clearly observed that the dislocations emit from the cutting region and some of them glide deep into the workpiece.

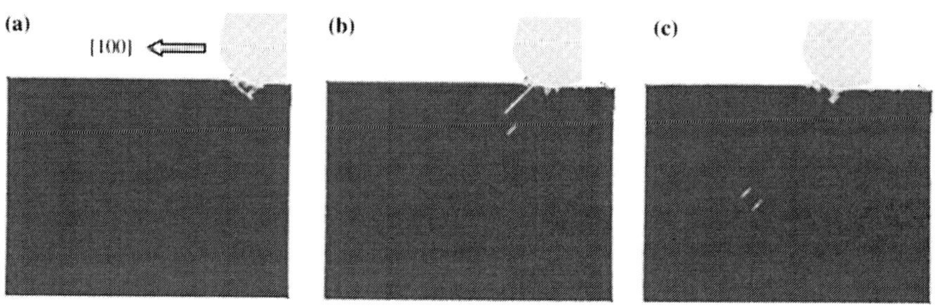

Figure 4. The simulated nanometric cutting process at the cutting distances of (**a**) 8 nm, (**b**) 12 nm and (**c**) 16 nm. The cutting depth is 0.8 nm and cutting speed is 100 m/s. The figures are shown in the cross-sectional views. The light blue color shows the cutting chips and dislocations inside the workpiece

Figure 5a–c present the side views of the 3D lattice defects at the cutting distances of 8, 12, and 16 nm, respectively. In the figures, the defect-free atoms in the workpiece are removed from the visualization. Note that the isolated atoms distributed inside the model are not lattice defects. They are left in the figures due to the thermal vibration of atoms. Although the CSP method is not perfect in identifying the lattice defects at finite temperature, it is more accurate than other methods such as the atomic coordinate number and the slip vector. It can be seen from Fig. 5a–c that lattice defects are formed in the workpiece during the cutting process. Moreover, a dislocation loop is generated and moves in the $[10\bar{1}]$ direction.

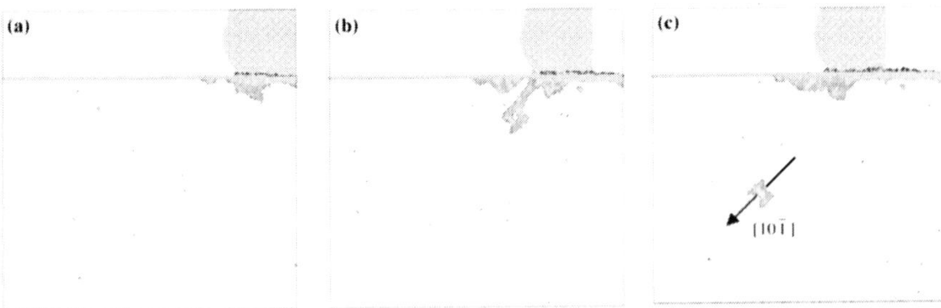

Figure 5. Side views of the lattice defects during the nanometric cutting process at the cutting distances of (**a**) 8 nm, (**b**) 12 nm and (**c**) 16 nm. The formation and movement of a dislocation loop inside the workpiece can be clearly seen

The cutting forces in the MD simulations are obtained by summing the atomic forces of all the workpiece atoms on the tool atoms. The variations of the cutting forces with the cutting distance during the cutting process for this simulation case are shown in Fig. 6. It can be seen that both the tangential cutting force,*Fx*, and the normal cutting force,*Fz*, increase at the start of the cutting. Then the cutting forces tend to remain steady during the rest of the cutting process. The formation of dislocations results in the release of the accumulated cutting energy, which corresponds to the temporary drop of the cutting force. The fluctuation of the cutting forces in Fig. 6is due to the formation of dislocations and their complex local movement in the cutting region. It is also observed from Fig. 6that the normal cutting force,*Fz*, shows stronger fluctuation than the tangential cutting force,*Fx*. This is because at this very small cutting depth, the normal cutting force is higher than the tangential cutting force, and therefore the normal cutting force undergoes stronger fluctuation. With a larger cutting depth as discussed in next section, the magnitude of normal cutting force is close to that of the tangential cutting force, and so magnitude of the force fluctuations is also close. The simulated cutting force in the thickness direction of the workpiece (*y* direction) is not shown here as it is very small with its time-averaged value over the whole cutting process being zero.

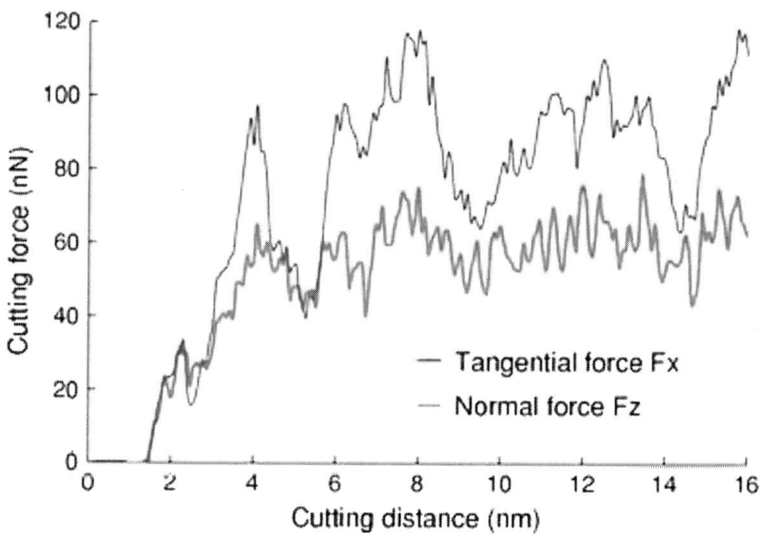

Figure 6. Variations of the cutting forces with the cutting distance

Effect of Cutting Depth on the Cutting Process

For nanometric cutting, it is interesting to understand how the cutting depth influences the cutting process. Figure7a–c show material deformation, chip formation, and dislocations during the cutting process for the cutting depths of 0.8, 2.0, and 4.0 nm, respectively. It can be seen that a larger cutting depth results in more workpiece material deformation around the tool and bigger cutting chip. Moreover, a larger cutting depth results in more lattice defects and dislocations in the cutting regions. However, the isolated dislocation loops is not observed in the cutting process for the cutting depth of 2.0 nm, though they are observed for the cutting depths of 0.8 and 4.0 nm. This shows that the dislocation activity is very complex in the nanoscale cutting process.

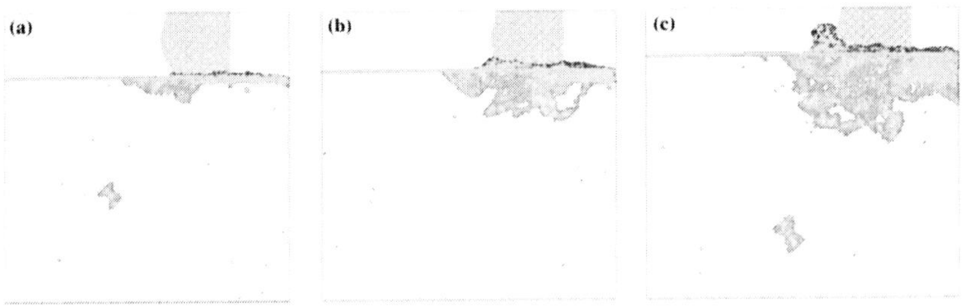

Figure 7. Material deformation and dislocations for the cutting depths of (**a**) 0.8 nm, (**b**) 2.0 nm and (**c**) 4.0 nm. The blue color shows the dislocations inside the workpieces

Figure 8 shows the time-averaged tangential and normal cutting forces during the cutting process for the different cutting depths. It can be seen that both tangential and normal cutting forces decrease as the cutting depth decreases. However, the tangential cutting force decreases faster than the normal cutting force. Consequently, the ratio of normal force to tangential force changes from smaller than 1.0 for the cutting depth of 4.0 nm to greater than 1.0 for the cutting depths of 2.0 and 0.8 nm. This shows that for nanoscale cutting with small cutting depth, as the tool edge radius is quite large compared to the cutting depth, the nanoscale cutting is more similar to the conventional grinding with a large negative tool rake angle.

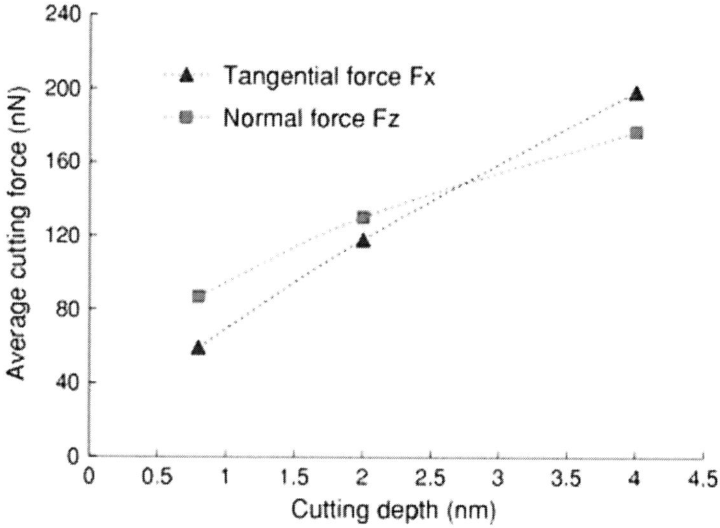

Figure 8. Variations of the time-average cutting forces with the cutting depth

Figure 9 shows the variations of the resultant cutting force and the specific cutting force with cutting depth. Here the resultant cutting force,*Fr*, is the vector sum of the tangential force,*Fx*, and normal force,*Fz*. Note that the average cutting force along the thickness direction*Fy* is zero. The specific cutting force,*Fs*, is the resultant cutting force divided by the cutting depth. It can be seen that with the decrease of cutting depth the resultant cutting force decreases. However, the specific cutting force increases rapidly with the decrease of cutting depth, which shows a very obvious "size effect". The "size effect" on the specific cutting force in nanometric cutting can be explained by the metallic bonding. The special feature of metallic bonding is that the strength of the individual bond has a strong dependence on the local environment. The bonding becomes stronger at the surface due to the localization of the electron density. The smaller the cutting depth, the larger the ratio of cutting surface to cutting volume, and thus the bigger the specific cutting force.

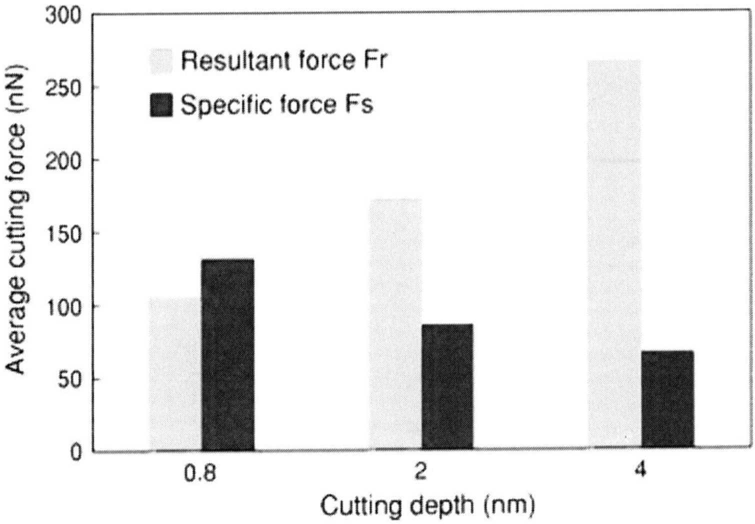

Figure 9. The resultant cutting force and the specific cutting force for the different cutting depths of 0.8, 2.0 and 4.0 nm

CONCLUSIONS

We have performed a series of large-scale 3D MD simulations using the EAM potential to study the nanometric cutting process. Three different model sizes of 2-million-atom, 4-million-atom, and 10-million-atom are used with different cutting speeds and cutting depths. It is shown that the 2-million-atom model, though quite large compared with the models used in the previously reported studies, is not large enough to eliminate the boundary effect for the simulation conditions used. It is also shown that the 4-million-atom model is large enough to eliminate the boundary effect at the cutting speed of 100 m/s and cutting depth of up to 4 nm. A detailed study on the material deformation, lattice defects, dislocation movement, and cutting forces during the cutting process is made with the 4-million-atom model. It is observed that the nanometric cutting process is accompanied by complex material deformation, chip formation, lattice defect generation, and dislocation movement. It is found that as the cutting depth decreases, both the tangential and normal cutting forces decreases; however, the tangential cutting force decreases faster than the normal cutting force. It is also found that as the cutting depth decreases, the specific cutting force increases, which reveals that the "size effect" exists in nanometric cutting.

ACKNOWLEDGMENTS

This work has been supported by the Agency for Science, Technology and Research (A*STAR), Singapore. Thanks also go to the staffs of the Computational Resource Centre at the Institute of High Performance Computing, who have provided the assistance in the large-scale computing and visualization.

REFERENCES

1. Masayoshi E, Takahito O: MEMS/NEMS by Micro Nanomachining. *IEIC Tech. Rep.* 2003, 103: 13.

2. Schumacher HW, keyser UF, Zeitler U, Haug RJ, Ebert K: Controlled mechanical AFM machining of two-dimensional electron systems: fabrication of a single-electron transistor. *Physica E* 2000

3. Maekawa K, Itoh A: Friction and tool wear in nano-scale machining–a molecular dynamics approach. *Wear* 1995

4. Zhang L, Tanaka H: Towards a deeper understanding of wear and friction on the atomic scale–a molecular dynamics analysis. *Wear* 1997

5. Komanduri R, Chandrasekaran N, Raff LM: MD simulation of nanoscale cutting of single crystal aluminum–effect of crystal orientation and direction of cutting. *Wear* 2000

6. Komanduri R, Chandrasekaran N, Raff LM: Molecular dynamics simulation of atomic-scale friction. *Phys. Rev. B* 2000

7. Komanduri R, Chandrasekaran N, Raff LM: MD simulation of exit failure in nanoscale cutting. *Mater. Sci. Eng. A* 2001, 311: 1. 10.1016/S0921-5093(01)00960-1

8. Fang TH, Weng CI: Three-dimensional molecular dynamics analysis of processing using a pin tool on the atomic scale. *Nanotechnology* 2000, 11: 148–153.

9. Zhang JJ, Sun T, Yan YD, Liang YC, Dong S: Molecular dynamics simulation of subsurface deformed layers in AFM-based nanometric cutting process. *Appl. Surf. Sci.* 2008, 254: 4774.

10. Daw MS, Baskes MI: Embedded-atom method: derivation and application to impurities, surfaces, and other defects in metals. *Phys. Rev. B* 1984, 29: 6443.

11. Finnis MW, Sinclair JE: A simple empirical N-body potential for transition metals. *Philos. Mag. A* 1984, 50:45.

12. Daw MS, Foiles SM, Baskes MI: The embedded-atom method: a review of theory and applications. *Mater. Sci. Rep.* 1993, 9: 251.

13. Pei QX, Lu C, Fang FZ, Wu H: Nanoscale cutting of copper: a molecular dynamics study. *Comput. Mater. Sci.* 2006, 37: 434.

14. Johnson RA: Analytic nearest-neighbor model for fcc metals. *Phys. Rev. B* 1988, 37: 3924.

15. Li J, Van Vliet KJ, Zhu T, Yip S, Suresh S: Atomistic mechanisms governing elastic limit and incipient plasticity in crystals. *Nature* 2002, 418: 307.

16. Zimmerman A, Kelchner CL, Hamilton JC, Foiles SM: Surface step effects on nanoindentation. *Phys. Rev. Lett.* 2001, 87: 165507.

17. Kelchner CL, Plimpton SJ, Hamilton JC: Dislocation nucleation and defect structure during surface indentation. *Phys. Rev. B* 1998, 58: 11085.

18. Kum O: Orientation effects of elastic-plastic deformation at surfaces: nanoindentation of nickel single crystals. *Mol. Simul.* 2005, 31: 115.

19. Lin YH, Jian SR, Lai YS, Yang PF: Molecular dynamics simulation of nanoindentation-induced mechanical deformation and phase transformation in monocrystalline silicon. *Nanoscale Res. Lett.* 2008,3: 71.

20. Tsuru T, Shibutani Y: Atomistic simulation of elastic deformation and dislocation nucleation in Al under indentation-induced stress distribution. *Model Simul. Mater. Sci. Eng.* 2006, 14: S55.

21. Saraev D, Miller RE: Atomic-scale simulations of nanoindentation-induced plasticity in copper crystals with nanoscale-sized nickel coatings. *Acta Mater.* 2006, 54: 33.

Chapter 2

Carbon Nanotubes and Graphene Nanoribbons: Potentials for Nanoscale Electrical Interconnects

Caterina Soldano [1,*], Saikat Talapatra [2] and Swastik Kar [3]

[1]*ETC srl, Bologna 40129, Italy*

[2]*Physics Department, Southern Illinois University at Carbondale, Carbondale, IL 62901, USA*

[3]*Department of Physics, Northeastern University, Boston, MA 02115, USA*

ABSTRACT

Carbon allotropes have generated much interest among different scientific communities due to their peculiar properties and potential applications in a variety of fields. Carbon nanotubes and more recently graphene have shown very interesting electrical properties along with the possibility of being grown and/or deposited at a desired location. In this Review, we will focus our attention on carbon-based nanostructures (in particular, carbon nanotubes and graphene nanoribbons) which could play an important role in the technological quest to replace copper/low-k for interconnect applications. We will provide the reader with a number of possible architectures, including single-wall as well as multi-wall carbon nanotubes, arranged in horizontal and vertical arrays, regarded as individual objects as well as bundles. Modification of their functional properties in order to fulfill interconnect applications requirements are also presented. Then, in the second part of the Review, recently discovered graphene and in particular graphene and few-graphene layers nanoribbons are introduced. Different architectures involving nanostructured carbon are presented and discussed in light of interconnect application in terms of length, chirality, edge configuration and more.

Keywords: carbon nanotubes; graphene nanoribbons; nanoscale interconnects

1. INTRODUCTION

Over the past few decades, the increasing progress in most technological areas from biomedicine to strategic deployment to meteorological forecasting has been largely benefitting from continuous and massive advances achieved in the field of computational technologies. With the need for larger quantities of data processing at faster pace within smaller volumes, the microelectronics industry has so far been steadily downsizing its basic active and passive building blocks. Those components are currently capable of operating at higher clock-speeds, delivering higher number of computations per second and, at the same time they occupy less physical space and consume less power. Nevertheless, the rapid progress towards gigascale miniaturization of those components faces a number of difficult challenges in terms of materials, architectures, fabrication and integration of nanoscale active and passive elements with expected performance in terms of reliability, speed, compatibility, and power consumption [1]. In this scenario, a fundamental issue of considerable difficulty is the development of the next generation interconnects for microprocessor units (MPUs).

Interconnects are wiring systems which distribute clock and other signals and provide power and ground, to and among the various circuits and systems functioning on a chip. As the various active and passive elements of the MPU reduce in dimension below 100 nm, also the corresponding interconnects will consistently follow. Existing interconnect technologies involving Cu/low-k lines and theirs evolutionary downsizing for sub-100 nm size interconnects face a number of important challenges. As the lateral dimension of interconnects approaches the mean free path of copper (~40 nm at room temperature [2]), the impact of grain boundary scattering, surface scattering, and the presence of a high-resistivity material as a diffusive barrier layer causes a rapid increase in the overall resistivity. Figure 1a shows how the resistivity of copper significantly and rapidly increases above its bulk value as interconnect lateral dimension shrinks below 100 nm [3], following which major degradation of reliability, signal transfer properties and thermal management issues start to become important. In addition, larger current densities for increasingly small interconnect lines will lead to an enhanced electro-migration failure. Tightly packed conducting material immersed in a dielectric environment will further introduce significantly larger parasitic capacitance, enhancing the delay of the signal propagation and limiting the overall operational bandwidth. This scenario clearly suggests that any length below 22 nm, which represents the current microelectronics community standard, copper will reveal to be inadequate for interconnect applications. Hence, a revolutionary new high-conducting material with higher failure current density and comparable or better performance metrics (such as specific capacitance and RC delay) is indeed in need for future copper replacement. Figure 1b represents the cross-sectional schematic of a typical Integrated Circuit (IC), reproduced from the International Technology Roadmap for Semiconductors (ITRS) map from 2009. In a typical IC, interconnect structures are scaled into local, intermediate, and global wiring levels based on optimization of delay, power, and bandwidth *etc.* and where each level has diverse requirements and at the same time poses different challenges. At the local (or M1) levels, resistivity and failure current density represent major bottlenecks, mostly in terms of delay. Global interconnects, which are the longest, will be impacted most by the degraded RC delay (*i.e.*, interconnect RC delay of 1892 ps for 1 mm Cu intermediate wire is thus not compatible). Significant effort has also been invested in tuning and modifying the dielectric properties of the interconnect material itself. Although the effective dielectric constant has been predicted to be progressively reduced (from 4 to 1.5) with developing technologies [4], nevertheless we have now reached a point in which no further important decrease is possible. Currently, the lowest value of the effective dielectric constant ranges between 2.5 and 3 and it has been stable over the last few years [5]. While some progress has been made in terms of evolutionary downsizing of existing technologies which will possibly work in the near-term (~22 nm), it is becoming increasingly evident that revolutionary new materials and/or processes are required to address long-term (<16 nm) scaling down of interconnects.

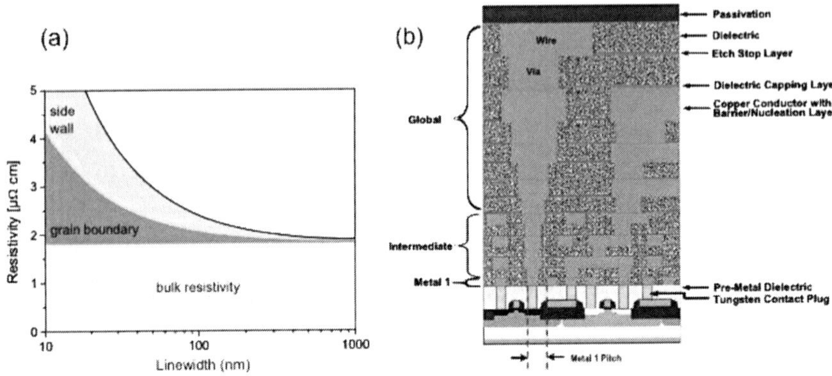

Figure 1. (**a**) Resistivity of copper as function of the interconnect linewidth; (**b**) Typical cross-section of hierarchical scaling (microprocessor unit (MPU) device) showing different interconnect scaling levels [3].

Different replacements for copper-based architectures have been proposed in the form of 3D Integrated Circuits, optical interconnects and carbon nanotubes (CNTs). Ongoing research in each of these field shows that those materials and structures could possibly successfully overcome the delay, power and bandwidth limitations of traditional interconnects, at one or more scaling levels. Nonetheless, in every single case, much remains to be done before one or more of these technologies can be realistically integrated with existing CMOS platforms while retaining their novel high-performance capabilities.

In this Review, we will focus our attention on carbon-based nanostructures (in particular carbon nanotubes and graphene nanoribbons) which could play an important role in the technological quest to replace copper/low-k for interconnect applications. We will provide the Reader with a number of possible architectures, including single-wall as well as multi-wall carbon nanotubes, arranged in horizontal and vertical arrays, and regarded as individual objects as well as bundles or arrays. Modification of their functional properties in order to fulfill interconnect applications requirements are also presented. Then, in the second part of the Review, recently discovered graphene and in particular graphene and few-graphene layers nanoribbons are introduced. Different architectures involving nanostructured carbon are presented and discussed in light of interconnect applications in terms of length, chirality, edge configuration and more.

We refer the Reader to more detailed literature regarding alternatives to carbon-based interconnects here presented (3D Integrated Circuit and Optical Interconnects) [6,7,8].

2. CARBON NANOTUBES

The ability to withstand current densities larger than 10^9 A/cm^2 (more than three orders of magnitude more than copper) even at ~1 nm diameter distinctly renders carbon nanotubes (CNTs) ideal candidates for nanoscale electrical interconnects [9]. CNTs are one-dimensional

(1D) quantum conductors that have been shown to have ballistic transport with mean free paths of the order of microns [10,11], which is significantly larger than copper. Carbon nanotubes can have diameters ranging from <1 nm (Single-Wall Nanotubes, SWNTs) up to tens of nanometers (Multi-Wall Nanotubes, MWNTs). While SWNTs can be either metallic or semiconducting, MWNTs are mostly metallic. In addition to their attractive ballistic conduction capabilities, these chemically and mechanically stable strands of carbon are extremely strong (tensile strength ~ 100 GPa) and possess very high thermal conductivity [12,13]. It has been shown that metallic SWNTs present values of the conductance close to their theoretically predicted values of 4 e^2/h and they are capable of functioning at tens of GHz range both as transistors and as interconnects [14,15]. Above everything else, CNTs have the natural geometrical shape for being a wire, *i.e.*, a nanometers-diameter cylinder with a large aspect ratio (it will be shown later on that it is easy to fabricate SWNTs that are microns long, and MWNTs that are centimeters long). All those favorable physical, electrical and geometric properties of carbon nanotubes make them an attractive and suitable choice as nanoscale electrical interconnects [16,17].

As compared to Cu/low-k interconnects, CNTs offer numerous advantages, as a results of the very strong nature of the C-C bonds, their 1D tubular shape and the electronic band structure. To benchmark Cu-based interconnects, carbon nanotubes have specific properties and fulfill the following requirements: (i) *reduced resistance*, in fact in the limit of short interconnects there exists a limitation associated to the quantum resistance of a single tube (for metallic SWNTs, the quantum of resistance is 6.5 kΩ [18], while for long interconnects the conductivity of nanotubes in densely-packed bundles is predicted to outperform copper counterpart; (ii) *resistance to electromigration*, due to the strong carbon-to-carbon bonds within the hexagonal lattice of the shell(s), resulting in the possibility by the tube to sustain elevated current density (10^9 A/cm^2), while for copper equivalent current density of 10^6 A/cm^2 already is capable of producing structural damages; (iii) large *thermal conductivity*, in fact isolated carbon nanotubes are predicted to have large values of thermal conductivity (few thousands W/mK [19,20]).

Increased resistance resulting when reducing size in *vias* structures can be addressed using bundles of carbon nanotubes; an additional degree of improvement can be obtained by increasing the site density. Further, since the energy gap of carbon nanotubes is inversely proportional to their diameters, it is important to control the growth of CNTs, which is always a mixture of metallic as well as semiconducting SWNTs, thus small-diameter SWNT must be reduced since will limit the conduction of the bundles and *vias* architectures.

2.1. Carbon Nanotubes for High-Density Integration

In nearly all calculations and most experiments, carbon nanotubes in the form of aligned arrays have been used to benchmark future interconnects against copper existing architectures. These structures could be of circular or rectangular cross sections, made of SWNTs or MWNTs bundles, and can be commonly treated as parallel nanotube architectures (PNAs). Those structures can potentially replace Cu/low-k interconnects at most levels of interconnect hierarchy except in places where low-resistance short interconnects are needed (*i.e.*, local interconnects), where CNT interconnects would be significantly more resistive than minimum-size copper wires since nanotubes as short as the gate pitch would be needed. In addition to the *classical* problems such as scattering associated with downsizing of copper interconnects, at the nanoscale, quantum confinement effects additionally play an important role. Recent computational predictions [21] have demonstrated that from a purely quantum-mechanical point of view, the resistance of carbon nanotube bundles is smaller than that of the copper wires for dimensions below 60 nm. Further, transmission line models [22] suggest that while ohmically contacted all-metallic SWNT bundle interconnects at local levels may be equivalent to copper in terms of RC delay, nevertheless the reduction in power dissipation and the increase of current density rendered them preferable even for the local level case. At the intermediate or global levels, RC delay becomes a crucial factor. In this case, larger diameter MWNT bundles, being chirality-independent all-metallic and having large mean free-paths, can achieve smaller signal delay than copper interconnects, with the improvements becoming more significant with technology scaling and increasing wire lengths. Saraswat *et al.* have predicted that for global levels, carbon nanotubes PNAs will outperform copper both in terms of latency as well as energy per bit of operation [23]. Monolayer or multilayer SWNTs PNAs can offer up to 50% reduction in capacitance and power dissipation with up to 20% improvement in latency if short enough (less than 20 μm) [24,25,26,27]. For semi-global interconnects, both latency and power dissipation can be substantially improved using bundles of SWNTs with reduced cross-sectional dimensions. From the electrical point of view, MWNT PNAs can outperform both SWNT and Cu at intermediate and global length scale in terms of conductivity. In addition, all-metallic SWNT PNAs appear to have much lower capacitance per unit length compared to copper and hence are ideal for local interconnects in terms of latency and power dissipation. This, in the long-term (sub-22 nm technology node) at all scaling levels, densely packed, all metallic PNAs, either SWNTs or MWNTs would be suitable copper replacements in terms of resistivity, failure current density, power dissipation, and delay. In the following sections, we will discuss some recent technological breakthroughs in terms of materials development, processing, structure fabrication, performance characterization, performance enhancement, and modeling of realistic SWNT and MWNT architectures that can potentially overcome the challenges faced by downward scaling of Cu/ low-k interconnects in terms of their projected performance metrics.

2.1.1. Scalable Ultra-High Density, Super Aligned SWNT PNAs

In order to compare or outperform Cu in terms of conductivity, it is essential to build highly aligned, short or/and long parallel arrays of

CNTs with very high number density. For 1 nm diameter SWNTs the ideal density for a densely-packed all-metallic bundle is 0.66 SWNTs/nm^2 has been proposed [1]. Jung *et al.* [28] have developed a template-based fluidic assembly process to build highly organized aligned single-wall nanotubes PNAs on SiO_2/Si substrate. Figure 2a schematically describes the steps involved in the fabrication of these structures. First, a plasma treatment is used to enhance the hydrophilic nature of SiO_2 surface, on which subsequently lithographically patterned photoresist is used to fabricate "channels". Substrates are then dip-coated into an appropriately functionalized SWNTs dispersion and the tubes selectively adhere to the SiO_2trenches. Removal of the resist leaves behind highly organized, dense, aligned arrays of SWNTs, the thickness of which (~tens of nanometers) can be controlled by the initial SWNT concentration and dipping speed. The dense filling factor and high degree of alignment of the structure is maintained end-to-end of the PNA structure, as shown in Figure 2b. The lateral size of the interconnect structures is limited only by the lithography process, with which sub-100 nm channels can be easily achieved. So-fabricated devices are characterized by an overall resistance ranging from few kΩ to tens of kΩ depending on the channel size (see Figure 2c), and they are able to withstand current density of ~10^7 A/cm^2. A very important and encouraging aspect of this technique is that the degree of alignment and the overall density appears to increase with smaller channel widths, which represents a great advantage in the perspective of interconnect applications. Further, the possibility of increasing the length in case of SWNTs by adding titanium to the catalyst surface leads to millimeter-long, small-diameter, single-wall carbon nanotube as a results of enhanced oxygen affinity [29]. Additional approach have also been proposed to fabricate SWNTs two-dimensional arrays such as electric field-assisted direct growth [30], low-feeding gas flow driven method [31] and AC dielectrophoresis (DEP) [32] from a solution of SWNTs, the last one leading to planar arrays of controllable density between pre-fabricated electrodes ranging from 0.5 SWNT/μm to more than 30 SWNT/μm, depending on the concentration of tubes in solution, as shown in Figure 2d,e. This approach leads to highly oriented PNAs with sheet resistance as low as few kΩ/□ depending on the density of carbon nanotubes (much lower compared to thin film of random nanotubes or nanotube cross bar junctions [33]) and the electrical properties of which can be tuned by the removal by breakdown of metallic SWNTs. Han and co-workers have demonstrated that the modification of the surface properties of single wall carbon nanotubes with short DNA linkers can favor the formation of parallel arrays where nanotubes are arranged one parallel to the others. In this case, DNA behaves as a spacer with a potential control of the inter-distance between adjacent tubes of less than 3 nm and the entire arrays can be easily transferred [34]. In principle this approach is very powerful (not limited to carbon nanotubes), however a post-fabrication densification step is in need to further increase the density of tubes.

2.1.2. Densely Packed Ultra-Long MWNTs

Multi-wall carbon nanotubes PNAs are ideally suitable for intermediate or global level interconnects since they are nearly all metallic due to larger diameters and they are characterized by μm-long mean free path. Each shell of properly contacted MWNTs can contribute to the conduction leading to low resistance conductive wires. Long MWNTs, with lengths ranging from 500 μm (intermediate) to 1000 μm (global), could potentially achieve significantly smaller signal delays [35] (as low as 15%) compared to copper equivalent. Various methods have been developed to grow very long and aligned MWNT bundles, with high yield and repeatability [36,37,38]. Li *et al.* have shown that ultra-long bundles can be grown in highly aligned parallel arrays of MWNTs, each nanotube running continuously end-to-end and that a selective control over the tube diameter (usually in the 25–40 nm range) can be obtained by adjusting the growth parameters [39]. Extraordinary growth can also be achieved through a multi-step *bottom-up* growth of self-supporting vertically aligned

MWNTs, which can easily reach mm long *Vias*-type structures with large degree of alignment [40], as shown in Figure 3.

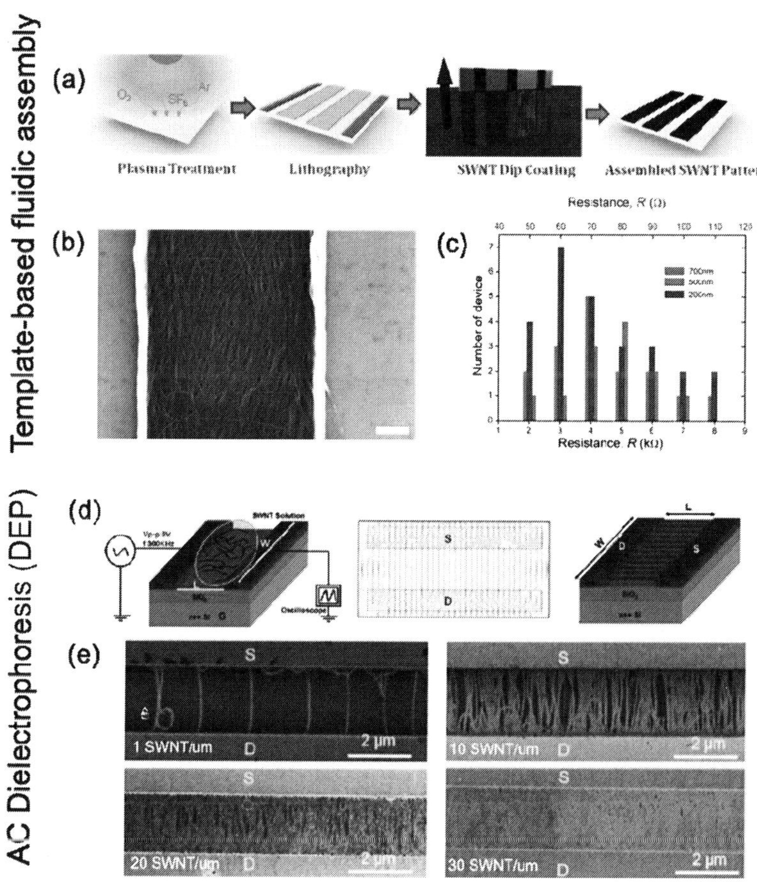

Figure 2. (**a**) Schematic of template-based fluidic assembly of high density, super aligned, sub-μm single-wall nanotube (SWNT) parallel nanotube architectures (PNAs); (**b**) High-magnification SEM images showing the degree of alignment of the SWNTs within channel widths of 1 μm. Scale bars = 200 nm; (**c**) Resistance histograms of SWNT-PNAs for different channel length (700, 500 and 200 nm, as labeled); (**d**) Schematic of the dielectrophoresis (DEP) assembly method; (**e**) SEM images for a typical DEP assembly with varying nanotube densities (1, 10, 20 and 30 tubes/μm). [28,32]

Figure 3. (Top) SEM image showing a two-dimensional array of pillars, each made of eight stacks of vertically aligned nanotube layers. Eight separate chemical vapor deposition (CVD) steps were used to grow the eight stack pillars on pre-patterned SiO_2. (Bottom) Higher-magnification SEM image showing the interfaces indicated by arrows between the separate stacks of nanotubes in a single pillar. The stacks are numbered 1 to 8, corresponding to the first to the eighth stack grown sequentially [40].

In addition, closely-packed carbon nanotube bundles are predicted to be characterized by higher conductivity when compared to copper; however, it is still extremely difficult to controllably grow those structures. Varying the feedstock flux during the CVD process (also known as "pulsed" CVD) can lead to dense array of all vertically aligned nanotubes: changing the gas flux and the duration of the pulse allows for the growth in predetermined length intervals, with precise density of the array over sub-μm lengths and tailored heights if growth can be reinitiated after each pulse [41]. Design and engineering of the catalyst offer the possibility of growing ultrahigh density, aligned forests of carbon nanotubes. Cyclic and repeated deposition/annealing/immobilization of catalyst thin films lead to nanotube forests with an area density of at least $10^{13}/cm^2$, over 1 order of magnitude higher than existing values, and very close to the limit of a fully dense forest [42]. Additional improvement (up to $10^{14}/cm^2$) can be reached if tube diameter can be further reduced. Post-growth methods have also been proposed in order to densify MWNTs arrays. Natural shrinking has been observed after immersion in solution due to simple solvent evaporation [43]. In as-grown aligned CNT arrays, the site density of nanotubes is about 60 CNTs/μm^2, orders of magnitude smaller than the effective maximum achievable site densities (~10^4 CNTs/μm^2). A solvent-assisted capillary coalescence method [44] is used to densify aligned CNT arrays in vertical pillars, shape very well suitable for vertical interconnect assemblies and vias, as shown in Figure 4.

The resultant pillars retain their cylindrical nature while uniformly shrinking in diameter, with site densities of approximately 900 CNTs/μm^2, with a densification factor of about 15. However, such structures will still present a significant unoccupied volume between individual CNT strands, and this could provide room for further densification. On the other hand, growth in templates (alumina, in most cases) allows for selective fabrication of all-vertically aligned and closely packed tubes with diameter as small as 30 nm [45]. Those structures, when still in template, could in principle allow for *all-walls top and bottom electrically contacted* nanotubes with uniform length and diameter. However, tubes grown in template are of poor quality when compared to arc-discharge tubes with a high level of defects which makes them highly resistive and, at the same time, the density (tube/μm^2) is very low.

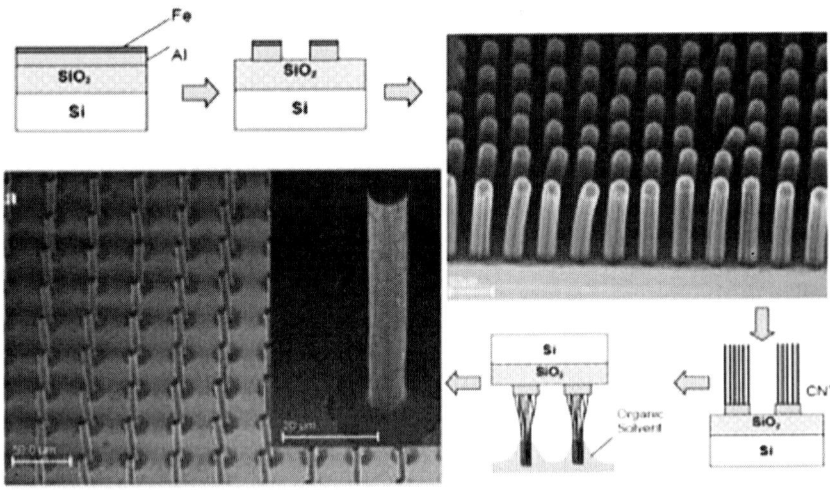

Figure 4. Schematics and SEM images describing capillary coalescence-driven densification using an organic solvent in aligned MWNT PNAs [44].

2.1.3. Selective Growth of Metallic Tubes

Statistically, about one third of as-grown SWNTs are metallic. These single shell tubes realistically represent the ultimate interconnects due their small radius and high conductance. However, there are no simple methods for selectively grow only metallic nanotubes, especially in more complex structures. Although great efforts have been put toward growth of controlled-diameter [46,47] and metallic SWNTs, very few reports have shown a certain degree of control over the growth. Yield of metallic SWNTs up to 65% of the final product has been achieved using as carbon source during the growth process monohydroxy alcohol homologues with a different ratio of carbon to oxygen content. It has been demonstrated that the yield of metallic

SWNTs can be improved by increasing the relative concentration of carbon to oxygen [48]. Further, Harutyunyan *et al.* were successful in achieving a fraction of metallic SWNTs of 91% by modifying the noble gas ambient during thermal annealing of the catalyst and in combination with oxidative and reductive species of the Fe nanocatalysts deposited on SiO_2/Si substrates. This work has clearly demonstrated the very strict connection between the catalysts morphology and the growing tube [49]. Modification of the catalyst properties (metallic phase and crystalline structure) can be a powerful and potential route in order to selectively tune the growth of carbon nanotubes with designated chirality [50,51]. Further on, detailed investigation of the effect of introducing iron pentacarbonyl ($Fe(CO)_5$) vapor during the growth process (arc-discharge) has shown the possibility of obtaining up to 90% of single wall metallic tubes [52]. Otherwise, it is possible to controllably grow double wall carbon nanotubes (DWNTs); in fact, due to their larger diameter, DWNTs have extremely small band-gaps, and at room or elevated temperatures, these tubes behave like near-zero gap metallic CNTs. Fu and coworkers have recently demonstrated the possibility of grow pillars on gold films of highly vertically aligned nanotubes, 80% of which are double wall carbon nanotubes [53], as shown in Figure 5. As compared to standard growth on Si substrate, it has been possible to considerably increase the number of DWNTs and less dispersive for other types of tubes. Water-assisted chemical vapor deposition on precisely-controlled thickness of iron catalytic films can be also applied to grow vertically aligned doubled-walled carbon nanotube forests with more than 2 mm in height with a yield of DWNTs approximately of 85% [54]. This synthesis approach also produces nearly catalyst-free DWNT forests with very high carbon purity (99.95%), which then potentially guarantees that the tube properties will not be affected by the presence of the catalyst. Koshio and coworkers have shown the possibility to obtain double-wall carbon nanotubes starting from highly dense small diameter MWNTs as a result of vacuum heat treatment at 2200 °C or higher, where, at these elevated temperature a selective removal of 1 to 3 shells within the MWNT takes place [55].

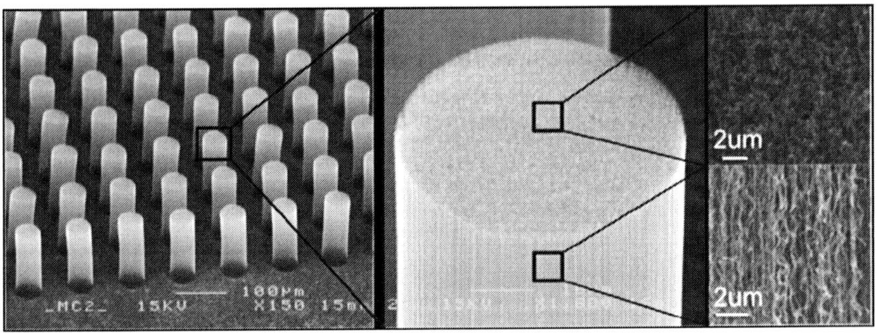

Figure 5. CNT bundles grown on the Au film, 80% of which are double wall carbon nanotubes (DWNTs). Top and side view show the large degree of vertical alignment of the tubes within the array [53].

Different post-growth approaches have also been proposed to selectively differentiate SWNTs, according to their chirality (and thus their electrical properties). An electric field-based treatment [56,57] can be used to separate all metallic tubes from semiconducting ones, leaving behind a parallel device array composed of a randomly grown mixture of metallic and semiconducting to reach densely packed devices in parallel where high density are required. In solution, single wall tubes can be well separated by density gradient ultracentrifugation (DGU), a technique traditionally used to separate and isolate different sub-cellular components, such as DNA and RNA. This method has been developed for sorting tubes based on chiral numbers and diameters, consisting in adding the SWNTs suspension in centrifuge tubes with varying density profile liquid mixtures; upon centrifugation, the tubes are separated and moved to regions (liquid) with matching densities [58,59,60]. Single-wall nanotubes can additionally be sorted using alternating current as a result of different dielectric constants depending on their electrical nature (metallic or semiconducting), as shown by Krupke and co-workers which have successfully separate in suspension metallic from semiconducting tubes [61].

2.1.4. Directional Growth of CNTs

Horizontal carbon nanotubes arrays represent still a challenge in the race for controllable and selective growth of carbon nanotubes. Interconnects applications are intrinsically in need of all the CNT to be of metallic nature. Being the energy gap of CNTs inversely proportional to the their diameter, the possibility of controlling the diameter during growth can allow for a control of the electronic properties of the tubes (and in particular their resistance [25,62,63]. There are three different approaches to achieve horizontal growth of carbon nanotubes: (i) electric-field based assembly/growth [64]; (ii) laminar gas flow [65] and (iii) step-templated growth on quartz/sapphire substrates [66,67]. Each of these methods unfortunately lead to very low nanotube densities or, as in the case of growth on quartz and sapphire, those are not compatible with growth on a chip. Further, one can think of first growing CNTs along the vertical direction and then proceed to flip them over.

Chai and coworkers have demonstrated that by controlling the catalyst thickness [68] on substrate (and the size of nanoparticles), it is possible to grow in the presence of an electric field bundles of MWNT with precise diameters along the horizontal direction [69]; in particular thicker catalyst films (5 nm) lead to large diameter tubes (25–35 nm) while in the case of thinner catalysts (1.2 nm), the diameter distribution of the grown tubes is around 5–7 nm. One other interesting outcome of this work is that catalyst thickness also allows for a certain control over the sites density of the CNTs bundles, where thinner catalyst leads to denser bundles.

Chiodarelli and coworkers have recently proposed a novel method to achieve horizontal growth of CNTs-based interconnects with lateral dimension comparable with current Cu equivalent [70]. CNTs are first grown vertically within pre-determined vias structures (diameters

200–300nm), then flipped horizontally and electrically contacted on both sides through metallization process. The as-fabricated interconnect structures then consist of carbon nanotube bundles with wall density of approximately $10^{13}/cm^2$, wire lengths of tens of μm and tube diameter of approximately 50 nm, characterized by nanotube resistivity as low as 1.1 mΩ·cm, value which is still larger than copper equivalent but which becomes closer and closer.

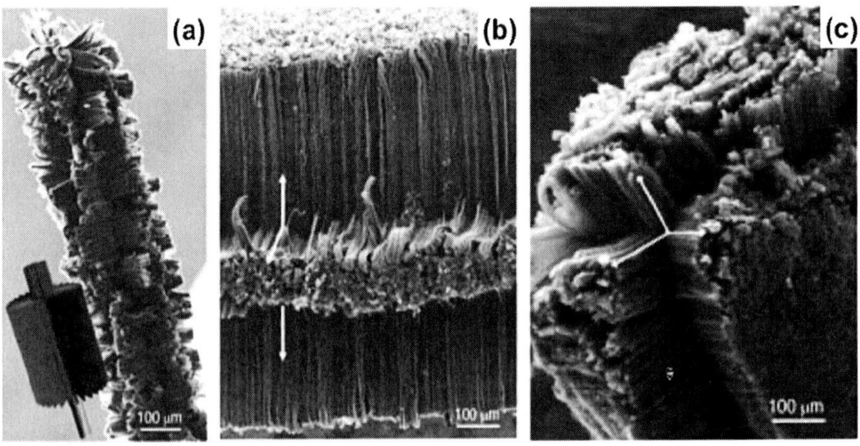

Figure 6. SEM of aligned MWNT arrays grown on metallic substrate showing growth directions conformal to the substrate orientation. Growth of carbon nanotubes on (**a**) metal wire, (**b,c**) metal sheets showing the highly directional nature of the CNT growth, with the orientation of the nanotubes being perpendicular to the substrate surface. The arrows indicate the direction of CNT growth [74].

The placement of catalysts on a vertical surface makes the horizontal growth harder to achieve than the vertical growth. Yan and coworkers have shown that a tilt evaporation technique can achieve catalyst patterning on vertical surfaces [71]. Upon demonstrating that $CoSi_2$ is a good catalyst support material in CNT interconnect application since it can be easily patterned on a vertical surface and support high density CNT growth, they have grown CNTs from the vertical surface with wall density of 6×10^{11} walls/cm^2 [72]. In case of single-wall and double-wall tubes, this challenge can be achieved using for example previously described template-based fluidic assembly methods, in case of MWNTs this can be achieved, to a large extent, by using floating vapor catalyst assisted CNT growth directly on metals [73]. Talapatra *et al*. have developed a standardized and unique technique for growing 3-D architectures of aligned bundles of carbon nanotubes on bulk metals, as shown inFigure 6 [74].

Growth on metal substrates has also the advantage of additionally providing an easy solution for growing electrically end-contacted carbon nanotubes. Parameters such as growth temperature and growth time can be selectively adjusted in order to obtain a specific tube length, as well in part their degree of alignment [75]. Electrically end-contacted dense bundles of aligned CNTs are promising candidates for interconnects and *via*structures at the sub-50-nm length scale. For optimal performance in these architectures, any metallization of individual tubes has to be on their ends, such that individual shell can be accessed by each electrode, and each nanotube in the bundle has to be individually metalized in this way. Post-growth metallization techniques are yet to reliably meet these criteria on a large scale. The unique advantage of the direct growth of CNTs on metallic substrates is that, within the array, every individual nanotube is electrically end-connected *in-situ*, overcoming several obstacles to applications in interconnect and contact technologies.

2.2. Carbon Nanotubes Devices and Architectures

2.2.1. Towards Low-Resistance Contacts

Large contact resistance represents still a technological challenge in order to realize interconnects. In the case of single-wall nanotubes PNAs, the sidewall resistance can be dramatically reduced using a cyclic high-current induced contact resistance annealing. This causes "hot-spots" at the contacts since that is where the resistance lies. The current density (~10^8–10^9 A/cm^2) is large enough to anneal the contacts, but causes no damage to the tubes. Electrical annealing treatment performed in a cyclic fashion on SWNT bundles of 0.5 μm in length, shows that the overall resistance change of more than 3 orders of magnitude due to contact annealing, bringing the total resistance below 1 kΩ. This method is potentially suitable for short SWNT PNAs. Rapid thermal annealing (RTA) at elevated temperatures (600–800 °C) for few tens of seconds can decrease the contact resistance between a carbon nanotube and metal electrodes (Ti/Au) by several orders of magnitude, becoming stable in the long-term as a result of a better adhesion of the carbon shell to the titanium film [76]. Welding at the nanoscale by a vibrating force at ultrasound frequency has been successful in contacting SWNTs onto metal electrodes, where the formed contact is mechanically robust and it is a stable low-Ohmic with resistances in the range of 8–24 kΩ (for a 1 μm long metallic single wall tubes at room temperature) [77]. For vertical integration, however different approaches have been considered such as pressure contacted soldering method [78], where lithographically patterned carbon nanotubes array grown on silica by chemical vapor deposition (CVD) are transferred of low-temperature (Ag/Sn) coated substrate. Figure 7a,b show the SEM images of carbon nanotubes arrays transferred onto solder pads (positive pattern) using the scheme described on the left as well as the CNTs transferred between the solder pads (negative pattern) in Figure 7c. Infiltration of CNT vias and subsequent polishing in order to expose open MWNT ends is also proposed for such types of architectures [44].

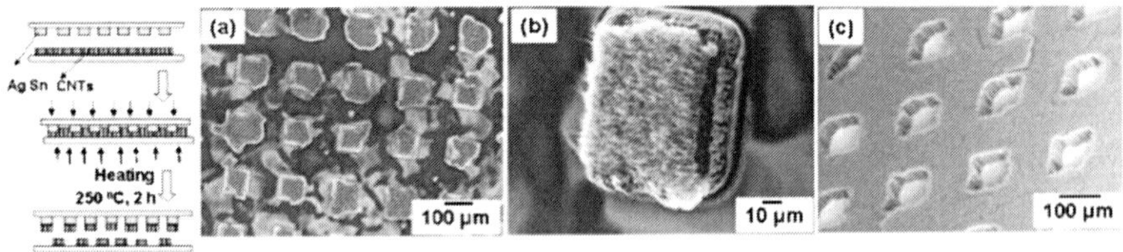

Figure 7. (**a,b**) SEM images of CNTs transferred onto solder pads (positive pattern) using the scheme described on the left; (**c**) CNTs transferred between the solder pads (negative pattern) [78].

Introduction of an interfacial carbon layer has also been proposed in order to reduce the contact resistance as a result of the improvement of the wetting capability and to the formation of chemical bonding to the tube [79]. Also, the contact area is larger due to the presence of the additional carbon layer.

2.2.2. Towards Defect-Free CNTS

Most high-quality CNTs are grown at extremely elevated temperatures (>1000 °C), thus resulting in carbon nanostructures with very low-defect densities. However, in carbon nanotubes grown at lower temperatures, post-growth methods are indeed in need for defects reduction. Ultrafast microwave annealing has been suggested as possible route towards the reduction of defect density in vertically aligned carbon nanotubes, for which only 3-minute-long exposure is necessary to induce a defect reduction, as suggested by Raman spectroscopy and thermogravimetric analysis. Mechanical as well as electric properties are also found to be enhanced upon microwave irradiation [80]. In the presence of a metal catalyst, defects formed during the addition of carbon atoms can be efficiently removed at the tube-catalyst interface, in a self-healing-type of approach. Simulations in fact show that potentially a defect-free CNT wall with 10^8–10^{11} carbon atoms can be achieved, thus leading to a potentially 0.1–100 cm long SWNT [81]. This effect is expected to take place through ring isomerization, leading to the removal of few (5–7) adatoms and monovacancy defects. The timescale of this process is of the order of several picoseconds and depended largely on the catalyst morphology and surface. This healing mechanism also potentially represents a way to control the chirality (n, m) of a growing SWNT during the growth process [82]. Among various catalysts, Fe has demonstrated to be the most efficient in defect healing [81].

Thermal annealing at high temperature has been proposed as an effective method to reduce defects in nanotubes [83,84,85,86]. Annealing at different high temperatures (>1000 °C) of CVD-grown multi-wall carbon nanotubes at relatively low temperatures (650 °C) in porous alumina template lead to a different degree of rearrangement of the imperfect graphitic structure and removal of weakly bounded defects, leading to the reconstruction of the hexagonal carbon lattice along the shell in a more ordered fashion. Transmission electron microscope (TEM) images in Figure 8a demonstrate the increase degree of graphitic order for increased temperatures.

Thermal reduction can also be used to purify and reduce chemical defects in the carbon lattice of single-wall carbon nanotubes, where a combination of first a high-temperature reactions in as-grown nanotubes in a pressurized chamber with a slow annealing process in vacuum environment is used [84]. Application of a rapid large bias across two terminals of a carbon nanotube not only decreases contact resistance, but also can dramatically reduce the number of defects. It has been shown that for highly defective MWNTs [87] (*i.e.*, grown in alumina template), a rapid cycling of the applied voltage across the electrodes of a device generates sufficient Joule heating to anneal defects in the CNT. This can drastically increase the mean free path (by 1–2 orders of magnitude) and reduce the resistance of tubes by orders of magnitude, as shown in Figure 8b [88]. Further, the investigation of the low-temperature charge transport properties allows estimating the defect density in those systems. Certain key transport parameters, such as number of quantum channels, disorder density (or equivalently, the mean free path), and *e-e* interactions play a strong role in determining the electronic properties of the interconnect itself. Using a combination of low-T transport measurement and application of field-theoretical models such as Lüttinger-liquid theory and Al'tshuler-Aronov [87] corrections can be used to estimate the number of quantum channels and mean free path in carbon nanotubes.

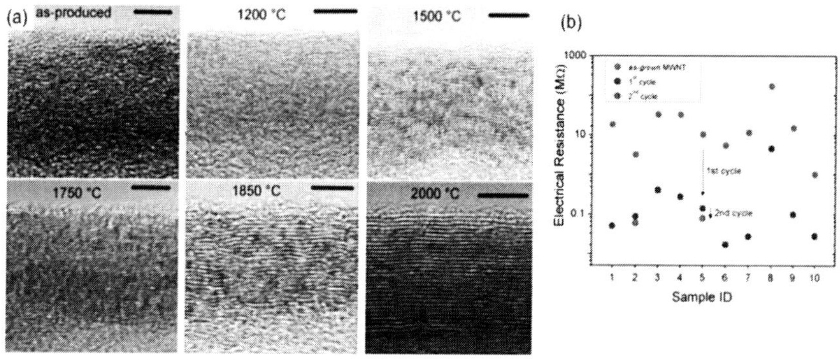

Figure 8. (**a**) Transmission electron micrographs of the walls of as-produced nanotubes grown in alumina template and at different annealing temperatures (as labeled). The initial, disordered carbon structure is converted into a more and more perfect graphitic structure with an increasing annealing temperature. Scale bar is 5 nm; (**b**) Variation of the electrical resistance in individual alumina template grown carbon nanotubes as function of the number of electrical cycles. [85]

2.2.3. Conductance Enhancement via Metal Decoration and Doping

Theoretical and experimental works have shown that modification of the structure of carbon nanotube can drastically affects the electronic properties. In particular, Density Functional Theory (DFT) calculations [89] show that decoration of SWNTs with platinum nanoclusters adds a number of electronic levels and enhances their density of states (DoS) near the Fermi level for both metallic and semiconducting carbon nanotubes, as shown inFigure 9a.

Figure 9b shows the effect of Pt nanocluster decoration on (top) the band-gap and (bottom) the conductance of two semiconducting ((8,0) and (9,0) and a metallic (10,0) single wall nanotube [28]. In case of semiconducting SWNTs, the addition of more than three atoms per cluster sizably brings down its band gap to values very close to zero, and remains close to zero for further decoration. In case of metallic SWNTs, the gap remains closed for values of n larger than 3. Further,

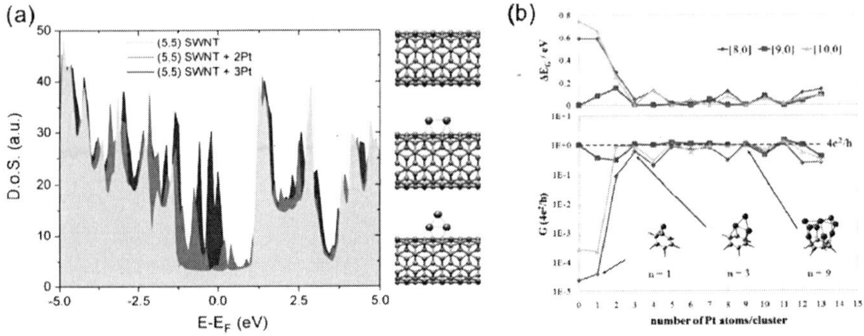

Figure 9. (**a**) Effect of platinum atom decoration on the (top) band-gap and (bottom) conductance of a metallic (9,0) and two semiconducting (8,0) and (10,0) SWNTs. For number of atoms > 3, all semiconducting SWNTs become metallic and their conductance approach the quantum limit of 4 e^2/h. Metallic tube remains metallic; (**b**) Cyclic high-bias treatment (HBT) treatment produced a Pt-nanoclusters decoration of the outer surface of the tube along with a conductance enhancement. (Inset) Platinum-decorated multi wall carbon nanotubes upon HBT [28].

calculation of conductance shows that in all three cases, platinum decoration causes the SWNT to approach or remain at the quantum limit of G~4 e^2/h, independently of the tube chirality. Calculations based on more rigorous GW approximation confirm metallic nature of Pt-decorated carbon nanotube [90]. Hence, decoration of carbon nanotubes by metal (Pt) clusters represents a general approach to convert semiconducting nanotubes into metallic, with a consequent effect of improving the overall conductivity of the SWNT PNAs. This process does not require neither selective growth nor post-growth selection of metallic SWNTs for interconnects applications. Encouraging results by electrochemical deposition of minute amounts of platinum on aligned SWNT PNAs have shown that those structures present negligible changes in surface morphology upon metal decoration and a dramatic decrease in resistivity of the order of approximately 75%. Experimentally, it has been demonstrated that decoration of platinum nanoclusters in MWNT has a significant enhancement of conductance. The process of metal decoration is achieved through a high-bias application procedure, which increases the number of quantum channels and decreases the defect density, and their combined effect can be extremely large (several orders of magnitude).

Chemical doping has also been proposed to enhance the conductivity of carbon nanotubes, both as individual objects as well as bundles [91,92,93]. The density of states (DoS) in doped carbon nanotubes reveals the possibility of increasing the conductivity of MWNTs; in fact, chemical doping of the outer shell has the overall effect of shift upwards the Fermi energy, leading to increased conductivity. Being the DoS independent from the tube diameter, elevated conductance values can be expected in case of metallic MWNTs rather than in metallic SWNTs as a results of both larger surface area available for doping and for the fact that any additional conduction channel is close to the Fermi energy in MWNTs. Doping with aryldiazonium and oxonium salts have been shown to be capable of reducing the resistance of MWNTs by almost 20% [94]. Lee *et al.* have shown that bulk samples of SWNTs grown by vapor phase reaction in the presence of bromine and potassium (each one as an example of *e*and *h*-type doping) can lead to an enhancement of the room temperature conductivity of a factor of about 30, which can outperform copper equivalent interconnects [95]. Doping multi wall carbon nanotubes with nitrogen (at low levels) and then performing thermal annealing at temperature of approximately 1000 °C, an elevated decrease in the electrical resistance of individual tubes is observed (about two orders of magnitude), where nitrogen is observed to be stable up to 1500 °C [96]. In case of SWNTs mats, increasing the dopant exposure (Cs) produces first a decrease and then an increase of their electrical resistance, allowing tunability of the dopant level to maximize conductivity of nanotubes structures [97]. In SWNTs mats, one or two order of magnitude increase in conductance can be achieved [95,97,98]. Conduction properties of potassium-doped individual ropes of SWNTs can also be modulated with a gate bias, further showing the intrinsic nature of *h*-doping in pristine tubes [99].

Zhao and coworkers have shown the possibility of fabricating iodine-doped, double-walled nanotube cables having electrical resistivity reaching ~10^{-7} Ω.m. The low-density and their specific density values outperform their copper counterpart and become comparable (one order of magnitude difference) with the highest specific conductivity metal such as sodium. In this case, elevated current handling capacity is also found (10^4–10^5 A/cm^2), along with no degradation of their electrical properties and freedom in terms of size and dimensions growth. Interestingly, a practical application in its preliminary form has been presented by using those doped-nanotubes cables in partly replacing metal connections in a commercially available light bulb circuit [100].

2.2.4. Breakdown in Carbon Nanotubes

One important aspect concerning CNT and their application as interconnects to be considered for realistic exploitation in ICs is the maximum current (density) the tube (or the array/bundle of tubes) can sustain before it breaks or partially loose their electrical properties. Carbon nanotubes both as individual object as well as in arrays/bundles can be tested for breakdown. It is important here to distinguish between single wall and multi wall carbon nanotubes, since the breaking of one single wall in SWNT will lead to the complete breakdown of the tubes, thus failure of the corresponding IC while in the case of MWNTs, there could be multiple and subsequent breakdowns accompanying the breaking of individual shells, leading to variation of the entire conduction properties prior to complete breakdown.

In most cases, breakdown in carbon nanotubes is induced by producing enough Joule heating to completely destroy the continuous carbon shell(s) or, if there are contacts, the contacts themselves. Joule heating can be produced thermally by direct heating and electrically by forcing a current along the tube(s) [101].

In case of single-wall carbon nanotube, electrical breakdown leads to an abrupt drop to zero of the current flowing within the tube, while in the case of multi wall carbon nanotubes and bundles exhibit higher current and multiple breakdown steps. Collins and coworkers have experimentally demonstrated that in arc-grown nanotubes, monitoring the current passing through the tube shows a staircase-shape where each step is associated with the progressive breakdown of adjacent and individual shells (12 μA/ shell) [102]. Similar experimental and theoretical results have been obtained on similar tubes grown by different techniques [18,103,104]. In some cases, the progressive removal of shell in carbon nanotubes, up to complete failure of the tubes, has been shown through electron microscopy (scanning and transmission) [105,106]. As mentioned, breakdown can also be induced thermally by simply heating the tubes. In air conditioning, due to presence of oxygen and self-heating, breakdown temperature is very well known of being approximately 600 °C. Hata and coworkers have carried out experiments both in vacuum (10^{-5} torr) and air and they observed that, in vacuum, carbon nanotubes breakdown takes place at much higher temperatures (approximately 1800 °C) [107]. This implies that oxidation cannot be the only cause leading to breakdown; different works have proposed an important role played by the presence of defects [108,109] and by the underlying substrate and its physical properties (*i.e.*, SiO_2 begin to melt at 1700 °C). Numerous studies have shown that the breakdown in carbon nanotubes is favored and initiated by the presence of oxygen (in air) which tends to etch the outermost shell [102], to the presence of defects along the wall where thus electric field (and consequently heat are intensified) [110,111], as well as to the poor thermal contact between CNTs and metal electrodes which prevents efficient heat removal [112]. It has been observed that metallic carbon nanotubes present the failure point at middle point location [113,114] (most likely corresponding to the hottest point) while semiconducting nanotubes breaks closer to one of the electrodes [111,115] (closer to drain), most likely corresponding to the area of the largest electric field and lower carrier density. Both these results indicate the diffusive nature of the heat dissipation and charge transport within the tube [116,117]. It has been demonstrated that a carbon nanotube grown by CVD with diameter of 18 nm can handle in vacuum a maximum current density up to 8×10^7 A/cm^2 with associated power of 0.6 mW (as compared to 0.16 mW and 2×10^7 A/cm^2 in air) [118]. Decoration with electrochemically deposited Ni particles allows protection of the CNT interconnect against oxidation and improvement of the heat release through the surrounding environment (about 1.5 mW before breaking down under vacuum, with a corresponding maximum current density as high as 1.2×10^8 A/cm^2). The Joule heating melts the nickel particles and promotes the formation of titanium carbon nitride with the overall results of improving the electrical contact between the CNT and the electrodes.

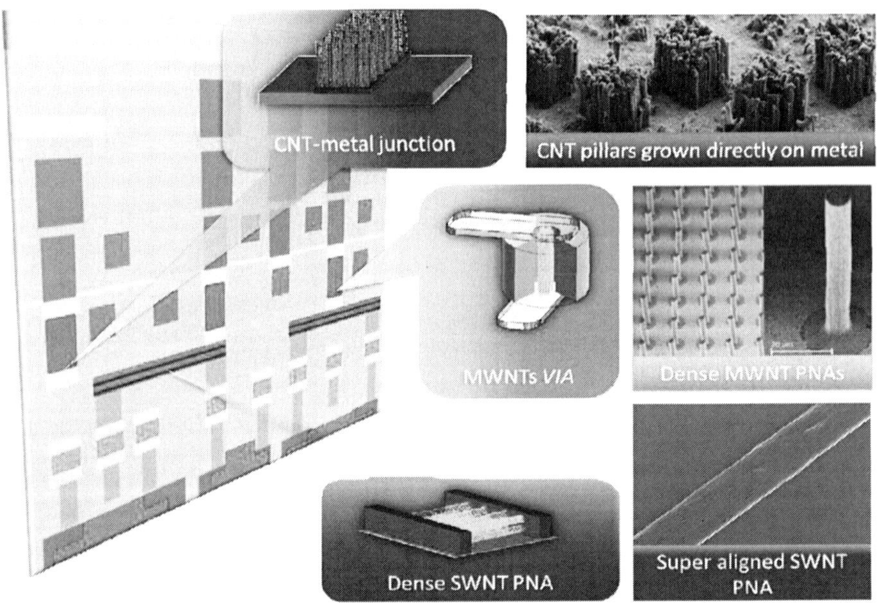

Figure 10. Schematic of a future Integrated Circuit (IC) chip with various interconnects potentially conceivable with CNT-based architectures, as described in the text which can hierarchically enable partial or full integrations in future IC chip [3,28,44,74].

The present Review has so far described various research efforts in the field of fabrication, characterization and development of ultra-high density, super-aligned nanotubes PNA-based high performance horizontal and vertical interconnect assemblies which can be possibly integrated into existing CMOS platforms. Figure 10 shows a typical cross section of an IC circuit (see Figure 1) and potential areas where various CNT-based architectures could begin to replace copper interconnect in the coming years. While it is important to fabricate high-performance CNT PNAs, one has to keep in mind the compatibility and integration considerations in terms of resistivity, failure current density, RC delay and operation bandwidth at appropriate individual scaling levels.

3. GRAPHENE AND GRAPHENE NANORIBBONS

3.1. Graphene as Two-Dimensional Crystal

Upon its experimental evidence in 2004, graphene has soon entered in the quest for the search of a new material for replacement for CMOS-compatible structures [3]. Graphene is a monolayer of *sp2*-bonded carbon atoms packed in a hexagonal lattice, arrangement that makes it rigorously a two-dimensional material [119]. Graphene has gathered the interest of the scientific and technological communities and, the investigation as well as implementation has been exponentially increasing due to the existing technology platform and knowledge, mostly borrowed from the decades-long carbon nanotube community. We refer the Reader to more specific literature for an in-depth and detailed description of the remarkable and interesting properties of this material [120,121]. In the present context, we will focus on some of the properties and characterization that are valuable and of great importance for interconnect applications.

Graphene is a zero bandgap semiconductor, in which electrons behaves as massless Dirac fermions and which could in principle offer less power consumption and enhanced performance due to increased carrier velocities. The ambipolar nature of graphene makes it a very versatile material since it can be used both as an *n*- and *p*-channel material. In fact, charge carriers have been observed to "puddle" in graphene layers at zero field so that both *n*- and *p*-carriers can co-exist within the same layer. Although graphene is a rather new material, a timeline for potential real applications in expected rather soon within the current decade (by 2019–2020 according to International Technology Roadmap for Semiconductors—Emerging Research Materials Section in 2009 [122]).

There exist a variety of methods to produce graphene, ranging from CVD-*like* methods or epitaxial growth on a silicon wafer [123], to mechanical exfoliation [124], chemical oxidation [125] or solvent exfoliation from highly oriented pyrolytic graphite (HOPG) [126,127], direct CVD epitaxy on single crystal metal substrates [128] and sublimation of silicon from SiC [129]. Recently proposed approaches such as growth on polymers (polyethylene terephtalhate, PET or polydimethylsiloxane, PDMS) further provide a new class of flexible and transparent substrates (mechanical bending and twist up to 11% without any loss in conductivity). So far, mechanical exfoliation of graphene has produced high quality films on silicon [119]. Nevertheless, this method is not currently capable of selective placement of graphene at a specific location, which makes it an inadequate route to device fabrication of integrated circuit technologies. Thickness control of several monolayers on a millimeter scale and carrier mobility of 25,000 cm^2/Vs have been demonstrated for grapheme grown on SiC (as a result of Si sublimation), where annealing temperature of at least 1200 °C in H_2 ambient condition are required [130]. Deposition of a thin layer of SiC [131] on a Si wafer followed by silicon evaporation has led to few-atomic layers thin graphene. A common drawback for the above-mentioned techniques is the high temperatures required during processing which might introduce additional defects in 300 mm and 450 mm wafers-line production or might not be compatible with some of the subsequent processes to be performed. Graphene can be grown on single crystal substrates (Ni [132], Ir [133], Pt [134]); however, this process has so far been revealed to be quite expensive. Continuous graphene films have been successfully grown by ambient pressure CVD on polycrystalline Ni which can then be subsequently transferred (by Ni wet-etching) on arbitrary substrates [128], producing films with sheet resistances as small as 280 Ω/□ and mobilities as high as 3700 cm^2/Vs. Large-scale graphene growth can be achieved on copper [135] with room temperature electron mobilities of more than 4000 cm^2/Vs. Post-growth transfer of those large-scale graphene domains represents a cheaper (compared to CVD) and effective route for real implementation into industrial standards.

Other techniques such as the solvent exfoliation [127] or the chemical route through production of graphene oxide are currently explored; firstly, graphene can be exfoliated in solvents from HOPG [126] with ultrasound, but solvent choices are very limited because of solvent surface energy thermodynamics. As for graphene oxide, this is a century-long well-established production method consisting first of oxidizing graphene (thermal or chemical intercalation), and then dispersing it in a solvent. However, graphene oxide shows fundamentally different electronic properties than individual graphene layer obtained by mechanical exfoliation or supported growth, thus requiring a chemical reduction step (in hydrazine or hydrogen plasma environment) to retrieve original electronic properties. Solution of chemically reduced grapheme oxide is found to consist primarily of monolayer graphene with mobility between 10 and 1,000 cm^2/Vs, with the conduction dominated by hopping transport through regions of highly ordered graphene surrounded by disordered grapheme [136].

Graphene elevated value of electron mobility at room temperature has rather quickly opened the way to the investigation of a variety of physical phenomenon, ranging from mesoscopic physics [137], Berry's phase [138] and unusual quantum Hall effect [139]. In the case of supported graphene devices, there is a fundamental upper limit of 40,000 cm^2/Vs [140] at room temperature due to trapped charges in the substrate itself [141]. Efficient screening [142] and/or complete removal of the underlying substrate are indeed required in order to enhance the mobility of graphene [143]. Further, it has been shown that in suspended and annealed graphene-based devices, the value of the mobility exceed 200,000 cm^2/Vs, which so far represents the highest value for a semiconductor or a semimetal [144,145]. The mobility is little affected by chemical doping [146] or high electric-field-induced carrier concentrations, suggesting the possibility of achieving rather easily ballistic transport regime on a sub-micrometer scale at 300 K [147].

3.2. Graphene and Few-Graphene Layers Nanoribbons

Although graphene behaves as a semi-metal, it has been shown that it is indeed possible to create an energy gap in graphene through different methods as (i) the inclusion of sp3 hydrocarbon defects in the sp2 lattice [148]; (ii) the distortion of the honeycomb lattice under uniaxial strain [149,150] and (iii) lateral confinement in *ribbon*-like shape [151]. While the first two approaches showing promising results are not seemingly likely to be implemented for interconnect applications, graphene nanoribbons could indeed represent valid and successful candidates [152]. It has been theoretically shown that the electronic properties of graphene nanoribbons can be tuned from perfectly metallic (zero bandgap) [153], in the case of zig-zag edge ribbons (ZGNR), to semiconducting for armchair ribbons (AGNR), where the engineered energy gap varies with the ribbon width and length. The origin of this energy gap in the band-structure of nanoribbons is attributed to edge roughness [151]. In addition, first principle calculations suggest that the carbon-carbon interatomic distance in the vicinity of the edges armchair GNRs is slightly smaller (~3.5% [154]) than the corresponding equivalent distance in two-dimensional graphene (1.42 Å), which can also induce the opening of an energy gap. This can also be favored if the spin component of zigzag GNRs is also considered. It has been recently shown that multilayer armchair GNRs are characterized by different bandgaps, depending on their width as a result of the interlayer interaction between adjacent layer which determines the magnetic polarization and band structures [155]. Various methods have been proposed to fabricate graphene nanoribbons. In particular, *e*-beam lithography allows producing nanoribbons with tailored lateral dimensions starting form graphene sheet, as shown in Figure 11 where the energy gap of NR varies as function of its width. However, these approaches are limited to a ribbon minimum width of approximately 20 nm, along with great difficulties in achieving smooth edges (roughness edges—larger than 5 nm—will affect the electronic properties of the ribbon itself) [151,156]. On the other hand, chemical routes leading to graphene oxide sheets and nanoplatelets can be followed for specific applications; nevertheless, reduction processes are needed to recover conductive properties of graphene [157,158], step that might not be compatible with interconnect technology process.

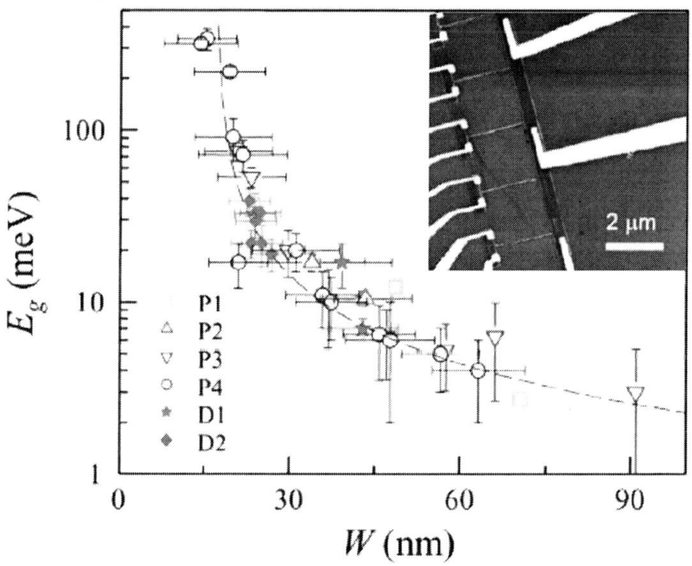

Figure 11. Energy gap dependence from ribbon width in lithographically fabricated graphene nanoribbons. Dashed lines in the inset show the value of as predicted by the empirical scaling of versus W. (Inset) SEM image of a representative graphene nanoribbon-based device [151].

Bottom-up approaches based on the molecular building-up of one-dimensional chains of polyaromatic carbon precursor have been developed [159]; even if this route surely represent a potentially extremely precise method to fabricate graphene nanotibbons with an accurate control over the edge roughness, it is indeed clear that its potential application to large-scale production are not yet ready.

Currently, solution-based methodologies are surely the most promising routes towards the production of graphene nanoribbons, where a very stable compromise between quantity and quality of the ribbons can be reached. Dai and coworkers [160] have shown a simple method to produce GNRs, starting from expandable graphite [161], as shown in Figure 12. Graphite is exfoliated by rapid heating at elevated temperature (1000 °C) in forming gas (Ar/H), dispersed in 1,2-dichloroethane (DCE) solution of poly(*m*-phenylenevinylene-co-2,5-dioctoxy-*p*-phenylenevinylene) and then sonicated. Once a homogeneous solution is obtained, centrifugation is used to separate large and heavy graphite pieces from light and narrow graphene small platelets. Ribbons with different widths down to sub-10 nm region can be thus produced. The same group has lately developed an unconventional and alternative approach based on unzipping mildly oxidized multi-walled carbon nanotubes through mechanical sonication. Quality of these ribbons in terms of disorder and edge roughness is very high, with conductance and mobility of 5 e^2/h and 1500 cm^2/Vs for 10–20 nm long ribbons [162].

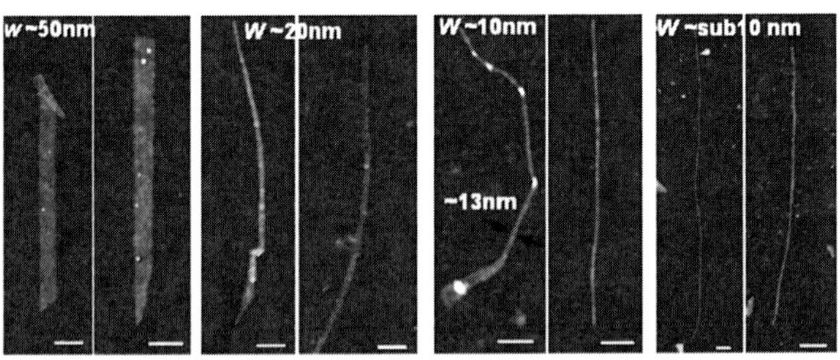

Figure 12. Atomic force microscopy images of selected chemically-derived Graphene Nanoribbons (GNRs) with widths of 50, 30, 20, 10 nm and sub-10-nm, respectively. Scale bar is 100 nm. [160]

Despite the fact that fully metallic GNRs might be hard to produce, this does not rule out the potential of GNRs as interconnects. In fact, semiconducting GNRs have a bandgap smaller than 0.1 eV, value that is equivalent to a copper interconnects with a diameter of 11 nm. If the Fermi energy is larger than half bandgap (few k_BT are sufficient), the first conduction band becomes adequately populated. Hence, for such GNRs, there would be a negligible difference between the ideal conductance of metallic and semiconducting GNRs [163]. Zig-zag and armchair nanoribbons (Fermi Energy = 0.2 eV, interlayer spacing = 0.35 nm) have significantly smaller resistivity than Cu equivalent and densely packed SWNT bundles. For the case of stack of non-interacting GNRs with smooth edges and Fermi energies above 0.2 eV has been predicted to outperform those of copper wires, especially at small cross-sectional dimensions and long lengths [162]. When compared to Cu/low-k interconnects, graphene nanoribbons hold promising potentials for the following reasons: (i) higher conductivity, with electron mean free path as large as a few hundred nanometer [130]; (ii) resistance to electro-migration due to the strong inherent carbon-carbon bonds which accounts for an extraordinary mechanical strength and a very large current conduction capacity of ~10^9 A/cm^2 (as compared to ~10^6 A/cm^2 in Cu) and (iii) larger in-plane thermal conductivity of approximately 5,300 W/Km, very much comparable to SWNTs bundles [164]. Nevertheless, there exist important current technological limitations and issues that prevent a much faster implementation for graphene nanoribbons in current interconnect applications.

Wafer-scale synthesis of high-quality graphene sheets on arbitrary substrates remains a major challenge. Successful small- and large-scale growths have been obtained on metals and on SiC substrates; however, the non-dielectric nature of the substrate (SiC) limits its applicability as interconnect. Much needed uniformity and grain size compatible with large-scale integration are indeed still in need. Then, to realistically implement the fabrication of GNRs, the crucial effect that edges play on the overall electronic properties of the GNR has to be taken into consideration. In fact, while the mean free path of non-patterned 2D graphene can be as large as several hundred nanometers, in the case of interaction between conducting electrons and edges, the effective mean free path becomes much smaller and hence, edge scattering becomes an a key issue in the electronic transport at the interconnect length scale. The mean free path associated with edge scattering depends on the edge quality (roughness), GNR width, and the ratio of transverse to longitudinal electron speed. Chemically derived GNRs have been shown to have smoother edges when compared to lithographically patterned ones, with a backscattering probability soon increasing from of 0.2, in the first case to 1 [165]. The product of GNR width and the ratio of longitudinal to transverse velocity determine the average length electrons move before interacting with edges [80]. In addition, doping could be used to shift the Fermi energy of graphene layers from the neutrality as a result of the charge accumulated at the graphene/substrate interface. However, this effect diminishes exponentially for upper layers because of the screening effect [166]. To use all the layers within a multilayer ribbon, the Fermi energy of upper layers must be shifted either through edge functionalization or doping, without any effect on the mean free path of electrons, otherwise, the conductance would become considerably smaller because of their bandgap. This conductance degradation is more severe for narrower GNRs with large bandgap.

In the case of GNRs, contact resistance still represent one of the key aspect/challenge in the real implementation of those structures into real interconnects. In fact, in a FET configuration, source-drain contacts have to be characterized by a low resistance contact, while maintaining unaltered the properties of the ribbon itself (*n*- or *p*-channel nature of the device). As in the case of charge transfer to generate doping, modeling has proposed the use of weak bonding contact metals to generate *n*-type contacts at work functions greater than 5.4 eV [167]. More recently, graphene nanoribbon edge states have been doped *n*-type through high temperature electrochemical ammonia treatment [168]. Experiments have demonstrated that *n*-type graphene can be produced through deposition of NH_3 and

CH_4 [169], while *p*-type graphene can be produced through deposition of H_2O and NO_2[170]. Experiments on graphene ribbons indicate that edges can convert to *p*-type, while the center of the ribbon remains *n*-type [171]. The challenge with these doping techniques will be to maintain the carrier doping in an integrated structure with interconnects. Ohmic contact formation may be easier than in small diameter carbon nanotubes, but a lot still has to be explored in this regard.

Once those numerous technological issues are addressed and overcome, GNRs could potentially be integrated for on-chip interconnect applications in the following two main categories:

1. *Graphite-like NRs* in place for lower interconnect resistance, especially for short interconnects. In the case of multiple layers, is indeed necessary that the layers in within the stack are yelectronically decoupled [81,120,172], condition which preserves graphene properties [173]. Further, in this case, all layers need to be electrically contacted.

2. *Few-layer GNR interconnects* suitable in order to reduce the interconnect capacitance (~ as much as 50%), delay and power dissipation of local interconnects. This arrangement is particularly interesting for short local interconnects in which the delay is dominated by capacitive loading and not resistivity.

4. CONCLUSIONS AND FUTURE CHALLENGES

Carbon-based nanostructures, and in particular carbon nanotubes and graphene nanoribbons, due to their numerous potential and outstanding properties, have rather quickly attracted the interest within different scientific communities in the quest towards replacement of presently used interconnects.

The present Review has provided the reader with a broad overview of such types of materials and architectures which might be suitable to replace current copper/low-k interconnects at various locations onto an integrated circuit. In particular, carbon nanotubes and graphene nanoribbons both hold great potentials at one or more interconnect levels. Although great advances have so far been achieved since the discovery of CNTs in the early 90s and more recently with the experimental evidence of graphene, many challenges and issues are still open regarding their complete exploitation in working devices. Carbon nanotubes and graphene nanoribbons have shown promising results in terms of RC delay, thermal management and conductance, both on individual objects as well as on assemblies in specific architectures; however, much remains to be advanced in order to achieve a full implementation of these architectures into commercially available IC in the next few years.

In fact, while moving towards the 16-nm technology node, crucial bottlenecks are now going to be faced. When downsizing elements below 100 nm size, interconnects face a number of important challenges. As the lateral dimension of interconnects approaches the mean free path of copper the impact of grain boundary scattering, surface scattering, and the presence of a high-resistivity material as a diffusive barrier layer causes a rapid increase in the overall resistivity. Thus, these phenomena cannot be neglected any further. A full integration of CNTs and GNRs in a real MPU requires the precise and strict control of material properties as well as processes to place those elements at desired locations. At the nanoscale, where at times only few hundreds/thousands of atoms are involved, it becomes of crucial importance to understand the nature of interfaces and any possible dynamics taking place at an interface. In this scenario in fact, metal contacts and electrodes on nanoscale objects behave differently when compared to micro- and macroscopic devices.

Precise control over the nature of defects and their density and location within carbon nanotubes and graphene nanoribbons is also significant and of key importance, since the electronic properties are very sensitive to the presence of any defect and their specificity. For example, controlling the edge roughness in the case of graphene ribbons is still an open and challenging issue that, if solved, might open the way to a massive use and exploitation of those materials within realistic architectures. Chirality engineering in carbon nanotubes and controlled growth of high-quality graphene at wafer-scale will surely be further needed towards the full implementation of these materials and their architectures in commercially available devices.

It is also important to note that in the last two decades (and more recently for GNRs), great advances have been achieved in placing CNTs/GNRs at precise location or within predefined arrays (with techniques such as dielectrophoresis or fluidic-based approaches); however, it is mandatory that those methods have to be compatible with any current MPU fabrication process. Further, any proposed materials, whether ultralong carbon nanotubes for long distance interconnects or graphene nanoribbons for short local interconnects, have to retain their properties throughout the entire MPU entire fabrication process.

Last, but not least, the economic impact of the introduction of these new materials in the large scale production of integrated circuits has to be considered. With confidence, the best materials to replace copper/low-*k*interconnects would be the ones that will surely bring innovation and improvement of the performances of both materials and devices. However, to reach a full implementation, at the same time, those materials and technologies will have to have less

ACKNOWLEDGEMENTS

CS would like to thank Italian MIUR funded project FIRB RBAP115AYN *Oxides at the nanoscale: multifunctionality and applications* for financial support. ST would like to acknowledge funding support from NSF-ECCS 0925682. SK would like to acknowledge financial support provided by NSF/ECCS-1102481.

REFERENCES

1. International Technology Roadmap for Semiconductors. *Interconnect Chapter*; SEMATECH: Austin, TX, USA, 2009, 2009 ed. Available online: http://www.itrs.net/Links/2009ITRS/2009Chapters_2009Tables/2009_Interconnect.pdf.
2. Wu, W.; Brongersma, S.H.; van Hove, M.; Maex, K. Influence of surface and grain-boundary scattering on the resistivity of copper in reduced dimensions. *Appl. Phys. Lett.* **2004**, *8*, 2838–1.
3. International Technology Roadmap for Semiconductors. *Interconnect Chapter*; SEMATECH: Austin, TX, USA, 2007, 2007 ed. Available online: http://www.itrs.net/Links/2007ITRS/2007_Chapters/2007_Interconnect.pdf.
4. International Technology Roadmap for Semiconductors, 1999 ed.; In *Interconnect Chapter*SEMATECH: Austin, TX, USA, 1999.
5. International Technology Roadmap for Semiconductors. *Interconnect Chapter*; SEMATECH: Austin, TX, USA, 2011, 2011 ed. Available online: http://www.itrs.net/Links/2011ITRS/2011Chapters/2011Interconnect.pdf.
6. Miller, A.B. Optical interconnects to silicon. *IEEE J. Sel. Top. Quantum Electron.* **2000**, *6*, 1312–1317.
7. Lu, D. Recent advances on chip-to-chip optical interconnects. *Proc. SPIE* **2009**.
8. Fischbach, R.; Lienig, J.; Meister, T. From 3D Circuit Technologies and Data Structures to Interconnect Prediction. In Proceeding of the 11th International Workshop on System Level Interconnect Prediction, San Francisco, CA, USA, 26–27 July 2009; pp. 77–84.
9. Wei, B.Q.; Vajtai, R.; Ajayan, P.M. Reliability and current carrying capacity of carbon nanotubes. *Appl. Phys. Lett.* **2001**, *79*, 1172–1174.

10. Charlier, J.-C.; Blase, X.; Roche, S. Electronic and transport properties of nanotubes. *Rev. Mod. Phys.* **2007**,*79*, 677–732.

11. Javey, A.; Guo, J.; Wang, Q.; Lundstrom, M.; Dai, H. Ballistic carbon nanotube field-effect transistors.*Nature* **2003**, *424*, 654–657.

12. Berber, S.; Kwon, Y.-K.; Tománek, D. Unusually high thermal conductivity of carbon nanotubes. *Phys. Rev. Lett.* **2000**, *84*, 4613–4616.

13. Choi, T.-Y.; Poulikakos, D.; Tharian, J.; Sennhauser, U. Measurement of the thermal conductivity of individual carbon nanotubes by the four-point three-method. *Nano Lett.* **2006**, *6*, 1589–1593.

14. Close, G.F.; Yasuda, S.; Paul, B.; Fujita, S.; Wong, P.H.S. A 1 GHz integrated circuit with carbon nanotube interconnects and silicon transistors. *Nano Lett.* **2008**, *8*, 706–709.

15. Chaste, J.; Lechner, L.; Morfin, P.; Fève, G.; Kontos, T.; Berroir, J.-M.; Glattli, D.C.; Happy, H.; Hakonen, P.; Plaçais, B. Single carbon nanotube transistor at GHz frequency. *Nano Lett.* **2008**, *8*, 525–528.

16. Xu, Y.; Srivastava, A. A model for carbon nanotube interconnects. *Int. J. Circuit Theory Appl.* **2009**, *38*, 559–575.

17. Srivastava, A.; Xu, Y.; Sharma, A.K. Carbon nanotubes for next generation very large scale integration interconnects. *J. Nanophotonics* **2010**, *4*, 041690:1–041690:27.

18. Frank, S.; Poncharal, P.; Wang, Z.L.; de Heer, W.A. Carbon nanotube quantum. *Resist. Sci.* **1998**, *280*, 1744–1746.

19. Van der Straten, O.; Zhy, Y.; Eisenbraun, E.; Kaloyeros, A. Thermal and Electrical Barrier Performance Testing of Ultrathin Atomic Layer Deposition Tantalum-Based Materials for Nanoscale Copper Metallization. In Proceedings of the

IEEE International Interconnect Technology Conference, Burlingame, CA, USA, 5 June 2002; pp. 188–190.

20. Watanabe, T.; Nasu, H.; Minamihaba, G.; Kurashima, N.; Gawase, A.; Shimada, M.; Yoshimizu, Y.; Uozumi, Y.; Shibata, H. Self-Formed Barrier Technology using CuMn Alloy Seed for Cu Dual-Damascene Interconnect with Porous-SiOC/ Porous-Par Hybrid Dielectric. In Proceedings of the IEEE International Interconnect Technology Conference, Burlingame, CA, USA, 4–6 June 2007; pp. 7–9.

21. Zhou, Y.; Sreekala, S.; Ajayan, P.M.; Nayak, S.K. Resistance of copper nanowires and comparison with carbon nanotube bundles for interconnect applications using first principles calculations. *J. Phys. Condens. Matter* **2008**, *20*, 095209:1–095209:9.

22. Maffucci, A.; Miano, G.; Villone, F. Performance comparison between metallic carbon nanotube and copper nano-interconnects. *IEEE Trans. Adv. Packag.* **2008**, *31*, 692–699.

23. Cho, H.; Koo, K.-H.; Kapur, P.; Saraswat, K.C. Performance comparisons between Cu/low-κ, carbon-nanotube, and optics for future on-chip interconnects. *IEEE Electron Device Lett.* **2008**, *29*, 122–124.

24. Naeemi, A.; Meindl, J.D. Design and performance modeling for single-walled carbon nanotubes as local, semiglobal, and global interconnects in gigascale integrated systems. *IEEE Trans. Electron Devices* **2007**, *54*, 26–37.

25. Naeemi, A.; Meindl, J.D. Compact physical models for multiwall carbon-nanotube interconnects. *IEEE Electron Device Lett.* **2006**, *27*, 338–340.

26. Naeemi, A.; Meindl, J.D. Monolayer metallic nanotube interconnects: Promising candidates for short local interconnects. *IEEE Electron Device Lett.* **2005**, *26*, 544–546.

27. Naeemi, A.; Meindl, J.D. Performance comparison between carbon nanotube and copper interconnects for gigascale integration (GSI). *IEEE Electron Device Lett.* **2005**, *26*, 84–86.

28. Kim, Y.L.; Li, B.; An, X.; Hahm, M.G.; Chen, L.; Washington, M.; Ajayan, P.M.; Nayak, S.K.; Busnaina, A.; Kar, S.; Jung, Y.J. Highly aligned scalable platinum-decorated single-wall carbon nanotube arrays for nanoscale electrical interconnects. *ACS Nano* **2009**, *3*, 2818–2826.

29. Ohashi, T.; Ochiai, T.; Tokune, T.; Kawarada, H. Increasing the length of a single-wall carbon nanotube forest by adding titanium to a catalytic substrate. *Carbon* **2013**, *57*, 79–87.

30. Peng, H.B.; Ristroph, T.G.; Schurmann, G.M.; King, G.M.; Yoon, J.; Narayanamurti, V.; Golovchencko, J.A. Patterned growth of single-walled carbon nanotube arrays from a vapor-deposited Fe catalyst. *Appl. Phys. Lett.* **2003**, *83*, 4238–4240.

31. Jin, Z.; Chu, H.; Wang, J.; Hong, J.; Tan, W.; Li, Y. Ultralow feeding gas flow guiding growth of large-scale horizontally aligned single-walled carbon nanotube arrays. *Nano Lett.* **2007**, *7*, 2073–2079.

32. Shekhar, S.; Stokes, P.; Khondaker, S.I. Ultrahigh density alignment of carbon nanotube arrays by dielectrophoresis. *ACS Nano* **2011**, *5*, 1739–146.

33. Kang, S.J.; Kocabas, C.; Kim, H.-S.; Cao, Q.; Meitl, M.A.; Khang, D.-Y.; Rogers, J.A. Printed multilayer superstructures of aligned single-walled carbon nanotubes for electronic applications. *Nano Lett.* **2007**, *7*, 3343–3348.

34. Han, S.-P.; Maune, H.T.; Barish, R.D.; Bockrath, M.; Goddard, W.A. DNA-inker-induced surface assembly of ultra dense parallel single walled carbon nanotube arrays. *Nano Lett.* **2012**, *12*, 1129–1135.

35. Hong, L.; Wen-Yan, Y.; Banerjee, K.; Jun-Fa, M. Circuit modeling and performance analysis of multi-walled carbon

nanotube interconnects. *IEEE Trans. Electron Devices* **2008**, *55*, 1328–1337.

36. Punbusayakul, N.; Ci, L.; Talapatra, S.; Surareungchai, M.; Ajayan, P.M. Ultralong aligned multi-walled carbon nanotube for electrochemical sensing. *J. Nanosci. Nanotechnol.* **2008**, *8*, 2085–2090.

37. Li, X.; Ci, L.; Kar, S.; Soldano, C.; Kilpatrick, S.J.; Ajayan, P.M. Densified aligned carbon nanotube films via vapor phase infiltration of carbon. *Carbon* **2007**, *45*, 847–851.

38. Hong, B.H.; Lee, J.Y.; Beetz, T.; Zhu, Y.; Kim, P.; Kim, K.S. Quasi-continuous growth of ultralong carbon nanotube arrays. *J. Am. Chem. Soc.* **2005**, *127*, 15336–15337.

39. Li, X.; Zhang, X.; Ci, L.; Shah, R.; Wolfe, C.; Kar, S.; Talapatra, S.; Ajayan, P.M. Air-assisted growth of ultra-long carbon nanotube bundles. *Nanotechnology* **2008**, *19*, 455609.

40. Li, X.; Cao, A.; Jung, Y.J.; Vajtai, R.; Ajayan, P.M. Bottom-up growth of carbon nanotube multilayers: Unprecedented growth. *Nano Lett.* **2005**, *5*, 1997–2000.

41. Jackson, J.J.; Puretzky, A.A.; More, K.L.; Rouleau, C.M.; Eres, G.; Geohegan, D.B. Pulsed growth of vertically aligned nanotube arrays with variable density. *ACS Nano* **2010**, *4*, 7573–7581.

42. Esconjauregui, S.; Fouquet, M.; Bayer, B.C.; Ducati, C.; Smajda, R.; Hofmann, S.; Robertson, J. Growth of ultrahigh density vertically aligned carbon nanotube forests for interconnects. *ACS Nano* **2010**, *4*, 7431–7436.

43. Engel, M.; Small, J.P.; Steiner, M.; Freitag, M.; Green, A.A.; Hersam, M.C.; Avouris, Ph. Thin film nanotube transistors based on self-assembled, aligned, semiconducting carbon nanotube arrays. *ACS Nano* **2008**, *2*, 2445–2452.

44. Liu, Z.; Ci, L.; Kar, S.; Ajayan, P.M.; Lu, J.-Q. Fabrication and electrical characterization of densified carbon nanotube micro-pillars for IC interconnection. *IEEE Trans. Nanotechnol.* **2009**, *8*, 196–203.

45. Phok, S.; Rajaputra, S.; Sing, V.P. Copper indium diselenide nanowire arrays by electrodeposition in porous alumina templates. *Nanotechnology* **2007**, *18*, 475601:1–475601:8.

46. Saito, T.; Ohshima, S.; Okazaki, T.; Ohmori, S.; Yumura, M.; Iijima, S. Selective diameter control of single-walled carbon nanotubes in the gas-phase synthesis. *J. Nanosci. Nanotechnol.* **2008**, *8*, 6153–6157.

47. Hiraoka, T.; Bandow, S.; Shinohara, H.; Iijima, S. Control on the diameter of single-walled carbon nanotubes by changing the pressure in floating catalyst CVD. *Carbon* **2006**, *44*, 1853–1859.

48. Wang, Y.; Liu, Y.; Li, X.; Cao, L.; Wei, D.; Zhang, H.; Shi, D.; Yu, G.; Kajiura, H.; Li, Y. Direct enrichment of metallic single-walled carbon nanotubes induced by the different molecular composition of monohydroxy alcohol homologues. *Small* **2007**, *3*, 1486–1490.

49. Harutyunyan, A.R.; Chen, G.; Paronyan, T.M.; Pigos, E.M.; Kuznetsov, O.A.; Hewaparakrama, K.; Kim, S.M.; Zakharov, D.; Stach, E.A.; Sumanasekera, G.U. Preferential growth of single-walled carbon nanotubes with metallic conductivity. *Science* **2009**, *326*, 116–120.

50. He, M.; Chernov, A.I.; Fedotov, P.V.; Obraztsova, E.D.; Rikkinen, E.; Zhu, Z.; Sainio, J.; Jiang, H.; Nasibulin, A.G.; Kauppinen, E.I.; Niemelä, M.; Krause, A.O.I. Selective growth of SWNTs on partially reduced monometallic cobalt catalyst. *Chem. Commun.* **2011**, *47*, 129–1221.

51. He, M.; Jiang, H.; Liu, B.; Fedotov, P.V.; Chernov, A.I.; Obraztsova, E.D.; Cavalca, F.; Wagner, J.B.; Hansen, T.W.;

Anoshkin, I.V.; *et al.* Chiral-selective growth of single-walled carbon nanotubes on lattice-mismatched epitaxial cobalt nanoparticles. *Sci. Rep.* **2013**.

52. Voggu, R.; Gosh, S.; Govindaraj, A.; Rao, C.N.R. New strategies for the enrichment of metallic single-walled carbon nanotubes. *J. Nanosci. Nanotechnol.* **2010**, *10*, 4102–4108.

53. Fu, Y.; Chen, S.; Bielecki, J.; Matic, A.; Wang, T.; Ye, L.-L.; Liu, J. Selective growth of double walled carbon nnaotubes on gold films. *Mater. Lett.* **2012**, *72*, 78–80.

54. Yamada, T.; Namai, T.; Hata, K.; Futaba, D.N.; Mizuno, K.; Fan, J.; Yudasaka, M.; Yumura, M.; Iijima, S. Size-selective growth of double-walled carbon nanotube forests from engineered iron catalysts. *Nat. Nanotechnol.* **2006**, *1*, 131–136.

55. Koshio, A.; Yudasaka, M.; Iijima, S. Disappearance of inner tubes and generation of double-wall carbon nanotubes from highly dense multiwall carbon nanotubes by heat treatment. *J. Phys. Chem. C* **2007**, *111*, 10–12.

56. Lastella, S.; Mallick, G.; Woo, R.; Karna, S.P.; Rider, D.A.; Manners, I.; Jung, Y.J.; Ryu, C.Y.; Ajayan, P.M. Parallel arrays of individually addressable single-walled carbon nanotube field-effect transistors. *J. Appl.Phys.* **2006**, *99*, 024302:1–024302:4.

57. Collins, P.G.; Arnold, M.S.; Avouris, P. Engineering carbon nanotubes and nanotube circuits using electrical breakdown. *Science* **2001**, *292*, 706–709.

58. Arnold, M.S.; Green, A.A.; Hulvat, J.F.; Stupp, S.I.; Hersam, M.C. Sorting carbon nanotubes by electronic structure using density differentiation. *Nat. Nanotechnol.* **2006**, *1*, 60–65.

59. Bonaccorso, F.; Hasan, T.; Tan, P.H.; Sciascia, C.; Privitera, G.; di Marco, G.; Gucciardi, P.G.; Ferrari, A.C. Density gradient ultracentrifugation of nanotubes: Interplay of bundling and

surfactants encapsulation. *J. Phys. Chem. C* **2010**, *114*, 17267–17285.

60. Ghosh, S.; Bachilo, S.M.; Weisman, R.B. Advanced sorting of single-walled carbon nanotubes by nonlinear density-gradient ultracentrifugation. *Nat. Nanotechnol.* **2010**, *5*, 443–450.

61. Krupke, R.; Hennrich, F.; von Lohneysen, H.; Kappes, M.M. Separation of metallic from semiconducting single-walled carbon nanotubes. *Science* **2003**, *301*, 344–347.

62. Nieuwoudt, A.; Massoud, Y. Evaluating the impact of resistance in carbon nanotube bundles for VLSI interconnect using diameter-dependent modeling techniques. *IEEE Trans. Electron Devices* **2006**, *53*, 2460–2466.

63. Su, Y.; hang, Y.; Wei, H.; Yang, Z.; Kong, E. S.-W.; Zhang, Y. Diameter-control of single-walled carbon nanotubes produced by magnetic field-assisted arc discharge. *Carbon* **2012**, *50,* 2556–2562.

64. Ural, A.; Li, Y.M.; Dai, H.J. Electric-field-aligned growth of single-walled carbon nanotubes on surfaces.*Appl. Phys. Lett.* **2002**, *81*, 3464:1–3464:3.

65. Huang, S.; Cai, X.; Liu, J. Growth of millimeter-long and horizontally aligned single-walled carbon nanotubes on flat substrates. *J. Am. Chem. Soc.* **2003**, *125*, 5636–5637.

66. Ismach, A.; Segev, L.; Wachtel, E.; Joselevich, E. Atomic-step-templated formation of single wall carbon nanotube patterns. *Angew. Chem.* **2004**, *116*, 6266–6269.

67. Kocabas, C.; Hur, S.H.; Gaur, A.; Meitl, M.A.; Shib, M.; Rogers, J.A. Guided growth of large-scale, horizontally aligned arrays of single-walled carbon nanotubes and their use in thin-film transistors. *Small***2005**, *1*, 1110–1116.

68. Wei, Y.Y.; Eres, G.; Merkulov, V.I.; Lowndes, D.H. Effect of catalyst film thickness on carbon nanotube growth by

selective area chemical vapor deposition. *Appl. Phys. Lett.* **2001**, *78*, 1394:1–1394:3.

69. Chai, Y.; Xiao, Z.; Chan, P.C.H. Horizontally aligned carbon nanotube bundles for interconnect application: Diameter-dependent contact resistance and mean free path. *Nanotechnology* **2010**, *21*, 235705.

70. Chiodarelli, N.; Fournier, A.; Okuno, H.; Dijon, J. Carbon nanotubes horizontal interconnects with end-bonded contacts, diameters down to 50 nm and lengths up to 20 μm. *Carbon* **2013**, *60*, 139–145.

71. Yan, F.; Zhang, C.; Cott, D.; Zhong, G.; Robertson, J. High-density growth of horizontally aligned carbon nanotubes for interconnects. *Phys. Status Solidi B* **2010**, *247*, 2669–2672.

72. Zhang, C.; Yan, F.; Allen, C.S.; Bayer, B.C.; Hofmann, S.; Hickey, B.J.; Cott, D.; Zhong, G.; Robertson, J. Growth of vertically-aligned carbon nanotube forests on conductive cobalt disilicide support. *J. Appl. Phys.* **2010**, *108*, 024311:1–024311:6.

73. Hiraoka, T.; Yamada, T.; Hata, K.; Futaba, D.N.; Kurachi, H.; Uemura, S.; Yumura, M.; Iijima, S. Synthesis of single- and double-walled carbon nanotube forests on conducting metal foils. *J. Am. Chem. Soc.* **2006**, *128*, 13338–13339.

74. Talapatra, S.; Kar, S.; Pal, S.; Vajtai, R.; Ci, L.; Victor, P.; Shaijumon, M.M.; Kaur, S.; Nalamasu, O.; Ajayan, P.M. Direct growth of aligned carbon nanotubes on bulk metals. *Nat. Nanotechnol.* **2006**, *1*, 112–116.

75. Pal, S.; Talapatra, S.; Kar, S.; Ci, L.; Vajtai, R.; Borca-Tasciuc, T.; Schadler, L.S.; Ajayan, P.M. Time and temperature dependence of multi-walled carbon nanotube growth on Inconel 600. *Nanotechnology* **2008**, *19*, 045610.

76. Lee, J.-O.; Park, C.; Kim, J.J.; Kim, J.; Park, J.W.; Yoo, K.H. Formation of low-resistance ohmic contacts between carbon

nanotube and metal electrodes by a rapid thermal annealing method. *J. Phys. D Appl. Phys.* **2000**, *33*, 1953–1956.

77. Chen, C.; Yan, L.; Kong, E.S.-W.; Zhang, Y. Ultrasonic nanowelding of carbon nanotubes to metal electrodes. *Nanotechnology* **2006**, *17*, 2192–2197.

78. Kumar, A.; Pushparaj, V.L.; Kar, S.; Nalamasu, O.; Ajayan, P.M.; Baskaran, R. Contact transfer of aligned carbon nanotube arrays onto conducting substrates. *Appl. Phys. Lett.* **2006**, *89*, 163120:1–163120:3.

79. Chai, Y.; Hazegi, A.; Takei, K.; Chen, H.-Y.; Chan, P.C.H.; Javey, A.A.; Wong, H.-S.-S.P. Low-resistance electrical contact to carbon nanotubes with graphitic interfacial layer. *IEEE Trans. Electron. Devices* **2012**, *59*, 12–19.

80. Lin, W.; Moon, K.-S.; Zhang, S.; Ding, Y.; Shang, J.; Chen, M.; Wong, C.-P. Microwave makes carbon nanotubes less defective. *ACS Nano* **2010**, *4*, 1716–1722.

81. Yuan, Q.; Xu, Z.; Yacobson, B.I.; Ding, F. Efficient defect healing in catalytic carbon nanotube growth. *Phys. Rev. Lett.* **2012**, *108*, 245505:1–245505:5.

82. Page, A.J.; Ohta, Y.; Okamoto, Y.; Irle, S.; Morokuma, K. Defect healing during single-walled carbon nanotube growth: A density-functional tight-binding molecular dynamics investigation. *J. Phys. Chem. C* **2009**, *113*, 20198–20207.

83. Tsai, J.T.H.; Tseng, A.A. Defect reduction of multi-walled carbon nanotubes by rapid vacuum arc annealing. *J. Exp.Nanosci.* **2009**, *4*, 87–93.

84. Tran, N.E.; Lambrakos, S.G. Purification and defect elimination of single-walled carbon nanotubes by the thermal reduction technique. *Nanotechnology* **2005**, *16*, 639–646.

85. Mattia, D.; Rossi, M.P.; Kim, B.M.; Korneva, G.; Bau, H.H.; Gogotsi, Y. Effect of graphitization on the wettability and

electrical conductivity of CVD-carbon nanotubes and films. *J. Phys. Chem. B* **2006**, *110*, 9850–9855.

86. Andrews, R.; Jacques, D.; Qian, D.; Dickey, E.C. Purification and structural annealing of multiwalled carbon nanotubes at graphitization temperatures. *Carbon* **2001**, *39*, 1681–1687.

87. Kar, S.; Soldano, C.; Chen, L.; Talapatra, S.; Vajtai, R.; Nayak, S.K.; Ajayan, P.M. Lüttinger iquid to Al'tshuler-Aronov transition in disordered, many-channel carbon nanotubes. *ACS Nano* **2009**, *3*, 207–212.

88. Soldano, C. Low-Temperature Electronic Transport in One-Dimensional Hybrid Systems: Metal Cluster Embedded Carbon Nanotubes. Ph.D. Thesis, Rensselaer Polytechnic Institute, Troy, NY, USA, 2007.

89. Parr, R.G.; Yang, W. *Density-Functional Theory of Atoms and Molecules*; Oxford University Press: Oxford, UK, 1989.

90. Hybertsen, M.S.; Louie, S.G. First-principles theory of quasiparticles: Calculation of band gaps in semiconductors and insulators. *Phys. Rev. Lett.* **1985**, *55*, 1418–1421.

91. Ryu, Y.; Yin, L.; Yu, C. Dramatic electrical conductivity improvement of carbon nanotube networks by simultaneous de-bundling and hole-doping with chlorosulfonic acid. *J. Mater.Chem.* **2012**, *22*, 6959–6964.

92. Zhou, C.; Wang, S.; Zhuang, Q.; Han, Z. Enhanced conductivity in polybenzoxazoles doped with carboxylated multi-walled carbon nanotubes. *Carbon* **2008**, *46*, 1232–1240.

93. Bockrath, M.; Hone, J.; Zettl, A.; McEuen, P.L.; Rinzler, A.G.; Smalley, R. Chemical doping of individual semiconducting carbon-nanotube ropes. *Phys. Rev. B* **2000**, *61*, R10606–R10608.

94. Neumann, A. Chemical Optimization of Carbon Nanotubes for Electronic Devices. Diploma Thesis, Technical University of Ilmenau, Ilmenau, Germany, 2005.

95. Lee, R.S.; Kim, H.J.; Fischer, J.E.; Thess, A.; Smalley, R.E. Conductivity enhancement in single-walled carbon nanotube bundles doped with K and Br. *Nature* **1997**, *388*, 255–257.

96. Fujisawa, K.; Tojo, T.; Muramatsu, H.; Elías, A.L.; Vega-Díaz, S.M.; Tristán-López, F.; Kim, J.H.; Hayashi, T.; Kim, Y.A.; Endo, M.; *et al.* Enhanced electrical conductivities of N-doped carbon nanotubes by controlled heat treatment. *Nanoscale* **2011**, *3*, 4359–4364.

97. Grigorian, L.; Sumanasekera, G.U.; Loper, A.L.; Fang, S.; Allen, J.L.; Eklund, P.C. Transport properties of alkali-metal-doped single-wall carbon nanotubes. *Phys. Rev. B* **1998**, *58*, R4195–R4198.

98. Ruzicka, B.; Degiorgi, L.; Gaal, R.; Thien-Nga, L.; Bacsa, R.; Salvetat, J.-P.; Forrò, L. Optical and dc conductivity study of potassium-doped single-walled carbon nanotube films. *Phys. Rev. B* **2000**, *61*, R2468–R2471.

99. Lee, R.S.; Kim, H.J.; Fischer, J.E.; Lefebvre, J.; Radosavljević, M.; Hone, J.; Johnson, A.T. Transport properties of a potassium-doped single-wall carbon nanotube rope. *Phys. Rev. B* **2000**, *61*, 4526–4529.

100. Zhao, Y.; Wei, J.; Vajtai, R.; Ajayan, P.M.; Barrera, E.V. Iodine doped carbon nanotube cables exceeding specific electrical conductivity of metals. *Sci. Rep.* **2011**, *1*, 83.

101. Mohsin, K.M.; Srivastava, A.; Sharma, A.K.; Mayberry, C. A thermal model for carbon nanotube interconnects. *Nanomaterials* **2013**, *3*, 229–241.

102. Collins, P.G.; Hersam, M.; Arnold, M.; Martel, R.; Avouris, P. Current saturation and electrical breakdown in multiwalled carbon nanotubes. *Phys. Rev. Lett.* **2001**, *86*, 3128–3131.

103. Collins, P.G.; Avouris, P. Multishell conduction in multiwalled carbon nanotubes. *Appl. Phys. A* **2002**, *74*, 329–332.

104. Dai, H.J.; Wong, E.W.; Lieber, C.M. Probing electrical transport in nanomaterials: Conductivity of individual carbon nanotubes. *Science* **1996**, *272*, 523–526.

105. Subramanian, A.; Dong, L.; Frutiger, D.; Nelson, B.J. Shell Engineering of Carbon Nanotube Arrays by Current Driven Breakdown. In IEEE-NANO 2006,In Proceedings of Sixth IEEE Conference on Nanotechnology, Cincinnati, OH, USA, 17–20 July 2006; Volume 2, pp. 901–904.

106. Lim, S.C.; Kim, K.S.; Lee, I.B.; Jeong, S.Y.; Cho, S.; Yoo, J.-E.; Lee, Y.H. Nanomanipulator-assisted fabrication and characterization of carbon nanotubes inside scanning electron microscope. *Micron* **2005**, *36*, 471–476.

107. Hata, K.; Futaba, D.N.; Mizuno, K.; Namai, T.; Yumura, M.; Iijima, S. Water-assisted highly efficient synthesis of impurity-free single-walled carbon nanotubes. *Science* **2004**, *306*, 1362–1364.

108. Huang, N.Y.; She, J.C.; Chen, J.; Deng, S.Z.; Xu, N.S.; Bishop, H.; Huq, S.E.; Wang, L.; Zhong, D.E.; Wang, E.G.; *et al.* Mechanism responsible for initiating carbon nanotube vacuum breakdown. *Phys. Rev. Lett.* **2004**,*93*, 075501:1–075501:4.

109. Yuzvinsky, T.D.; Mickelson, W.; Aloni, S.; Begtrup, G.E.; Kis, A.; Zettl, A. Shrinking a carbon nanotube.*Nano Lett.* **2006**, *6*, 2718–2722.

110. Molhave, K.; Gudnason, S.B.; Pedersen, A.T.; Clausen, C.H.; Horsewell, A.; Boggild, P. Transmission electron microscopy study of individual carbon nanotube breakdown caused by Joule heating in air. *Nano Lett.* **2006**, *6*, 1663–1668.

111. Berger, C.; Yi, Y.; Gezo, J.; Poncharal, P.; de Heer, W.A. Contacts, non-linear transport effects and failure in multi-walled carbon nanotubes. *New J. Phys.* **2003**, *5*, 1–16.

112. Pop, E. Energy dissipation and transport in nanoscale devices. *Nano Res.* **2010**, *3*, 147–169.

113. Lee, Y.H.; Lee, J.H.; Chung, S.J.; Lee, S.; Ju, B.K. Carrier carrying capacity of one-step grown suspended carbon nanotube bridge with carbon nanotube contact electrodes: For practical one-dimensional electronics. *Appl. Phys. Lett.* **2006**, *89*, 07310:1–07310:3.

114. Yuzvinsky, T.D.; Mickelson, W.; Aloni, S.; Konsek, S.L.; Fennimore, A.M.; Begtrup, G.E., Kis; Regan, B.C.; Zettl, A. Imaging the life story of nanotube devices. *Appl. Phys. Lett.* **2005**, *87*, 083103:1–083103:3.

115. Wei, Y.; Liu, P.; Jiang, K.L.; Liu, L.; Fan, S.S. Breaking single-walled carbon nanotube bundles by Joule heating. *Appl. Phys. Lett.* **2008**, *93*, 023118:1–023118:3.

116. Pop, E.; Mann, D.A.; Goodson, K.E.; Dai, H.J. Electrical and thermal transport in metallic single-wall carbon nanotubes on insulating substrates. *J. Appl. Phys.* **2007**, *101*, 093710:1–093710:10.

117. Pop, E.; Mann, D.; Wang, Q.; Goodson, K.; Dai, H.J. Thermal conductance of an individual single-wall carbon nanotube above room temperature. *Nano Lett.* **2006**, *6*, 96–100.

118. Santini, C.A.; Vereecken, P.M.; Volodin, A.; Groeseneken, G.; de Gendt, S.; van Haesendonck, C. A study of Joule heating-induced breakdown of carbon nanotube interconnects. *Nanotechnology* **2011**, *22*, 395202:1–395202:9.

119. Geim, A.K.; Novoselov, K.S. The rise of graphene. *Nat. Mater.* **2007**, *6*, 183–191.

120. Soldano, C.; Mahmood, A.; Dujardin, E. Production, properties and potential of graphene. *Carbon* **2010**, *48*, 2127–2150.

121. Zhu, Y.; Murali, S.; Cai, W.; Li, X.; Suk, J.W.; Potts, J.R.; Ruoff, R.S. Graphene and graphene oxide: Synthesis, properties, and applications. *Adv. Mater.* **2010**, *22*, 3906–3924.

122. International Technology Roadmap for Semiconductors. *Emerging Research Materials Chapter*; SEMATECH: Austin, TX, USA, 2009, 2009 ed. Available online: http://www.itrs.net/Links/2009Chapters_2009Tables/2009_ERM.pdf.

123. Kim, K.S.; Zhao, Y.; Jang, H.; Lee, S.Y.; Kim, J.M.; Kim, K.S.; Ahn, J.-H.; Kim, P.; Choi, J.-Y.; Hong, B.H. Large-scale pattern growth of graphene films for stretchable transparent electrodes. *Nature* **2009**, *457*, 706–710.

124. Novoselov, K.S.; Geim, A.K.; Morozov, S.V.; Jiang, D.; Zhang, Y.; Dubonos, S.V.; Grigorieva, I.V.; Firsov, A.A. Electric field effect in atomically thin carbon films. *Science* **2004**, *306*, 666–669.

125. Brodie, B.C. On the atomic weight of graphite. *Philos. Trans. R. Soc. Lond.* **1859**, *149*, 249–259.

126. Hernandez, Y.; Nicolosi, V.; Lotya, M.; Blighe, F.M.; Sun, Z.; De, S.; McGovern, I.T.; Holland, B.; Byrne, M.; Gun'Ko, Y.K.; *et al.* High-yield production of graphene by liquid-phase exfoliation of graphite. *Nat. Nanotechnol.* **2008**, *3*, 563–568.

127. An, X.; Simmons, T.; Shah, R.; Wolfe, C.; Lewis, K.M.; Washington, M.; Nayak, S.K.; Talapatra, S.; Kar, S. Stable aqueous dispersions of noncovalently functionalized graphene from graphite and their multifunctional high-performance applications. *Nano Lett.* **2010**, *10*, 4295–4301.

128. Reina, A.; Jia, X.; Ho, J.; Nezich, D.; Son, H.; Bulovic, V.; Dresselhaus, M.S.; Kong, J. Large area, few-layer graphene films on arbitrary substrates by chemical vapor deposition. *Nano Lett.* **2009**, *9*, 30–35.

129. Berger, C.; Song, Z.M.; Li, T.B.; Li, X.B.; Ogbazghi, A.Y.; Feng, R.; Dai, Z.; Marchenkov, A.N.; Conrad, E.H.; First, P.N.; *et al.* Ultrathin epitaxial graphite: 2D electron gas properties and a route toward graphene-based nanoelectronics. *J. Phys. Chem. B* **2004**, *108*, 19912–19916.

130. Berger, C.; Song, Z.; Li, X.; Wu, X.; Brown, N.; Naud, C.; Mayou, D.; Li, T.; Hass, J.; Marchenkov, A.N.; *et al.* Electronic confinement and coherence in patterned epitaxial graphene. *Science* **2006**, *312*, 1191–1196.

131. Suemitsu, M.; Miyamoto, Y.; Handa, H.; Konno, A. Graphene formation on a 3C-SiC(111) thin film grown on Si(110) substrate. *e-J. Surf. Sci. Nanotechnol.* **2009**, *7*, 311–313.

132. Gamo, Y.; Nagashima, A.; Wakabayashi, M.; Terai, M.; Oshima, C. Atomic structure of monolayer graphite formed on Ni(111). *Surf. Sci.* **1997**, *374*, 61–64.

133. Coraux, J.; N'Diaye, A.T.; Engler, M.; Busse, C.; Wall, D.; Buckanie, N.; Meyer zu Heringdorf, F.J.; van Gastel, R.; Poelsema, B.; Michely, T. Growth of graphene on Ir(111). *New J. Phys.* **2009**, *11*, 023006:1–023006:25.

134. Preobrajenski, A.B.; Ling Ng, M.; Vinogradov, A.S.; Mårtensson, N. Controlling graphene corrugation on lattice-mismatched substrates. *Phys. Rev. B* **2008**, *78*, 073401.

135. Li, X.; Cai, W.; An, J.; Kim, S.; Nah, J.; Yang, D.; Piner, R.; Velamakanni, A.; Jung, I.; Tutuc, E.; *et al.* Large area synthesis of high-quality and uniform graphene films on copper foils. *Science* **2009**, *324*, 1312–1314.

136. Gómez-Navarro, C.; Weitz, R.T.; Bittner, A.M.; Scolari, M.; Mews, A.; Burghard, M.; Kern, K. Electronic transport properties of individual chemically reduced graphene oxide sheets. *Nano Lett.* **2007**, *7*, 3499–3503.

137. Novoselov, K.S.; Geim, A.K.; Morozov, S.V.; Jiang, D.; Katsnelson, M.I.; Grigorieva, I.V.; Dubonos, S.V.; Firsov, A.A.

Two-dimensional gas of massless Dirac fermions in graphene. *Nature* **2005**, *438*, 197–200.

138. Zhang, Y.B.; Tan, Y.W.; Stormer, H.L.; Kim, P. Experimental observation of the quantum Hall effect and Berry's phase in graphene. *Nature* **2005**, *438*, 201–204.

139. Ozyilmaz, B.; Jarillo-Herrero, P.; Efetov, D.; Abanin, D.A.; Levitov, L.S.; Kim, P. Electronic transport and quantum hall effect in bipolar graphene p-n-p junctions. *Phys. Rev. Lett.* **2007**, *99*, 166804:1–166804:4.

140. Chen, J.H.; Jang, C.; Adam, S.; Fuhrer, M.S.; Williams, E.D.; Ishigami, M. Charged- impurity scattering in graphene. *Nat. Phys.* **2008**, *4*, 377–381.

141. Chen, J.H.; Jang, C.; Xiao, S.D.; Ishigami, M.; Fuhrer, M.S. Intrinsic and extrinsic performance limits of graphene devices on SiO_2. *Nat. Nanotechnol.* **2008**, *3*, 206–209.

142. Lin, Y.M.; Avouris, P. Strong suppression of electrical noise in bilayer graphene nanodevices. *Nano Lett.***2008**, *8*, 2119–2125.

143. Bolotin, K.I.; Sikes, K.J.; Jiang, Z.; Klima, M.; Fudenberg, G.; Hone, J.; Kim, P.; Stormer, H.L. Ultrahigh electron mobility in suspended graphene. *Solid State Communications* **2008**, *146*, 351–355.

144. Bolotin, K.I.; Sikes, K.J.; Hone, J.; Stormer, H.L.; Kim, P. Temperature dependent transport in suspended graphene. *Phys. Rev. Lett.* **2008**, *101*, 096802:1–096802:4.

145. Du, X.; Skachko, I.; Barker, A.; Andrei, E.Y. Approaching ballistic transport in suspended graphene. *Nat. Nanotechnol.* **2008**, *3*, 491–495.

146. Schedin, F.; Geim, A.K.; Morozov, S.V.; Hill, E.W.; Blake, P.; Katsnelson, M.I.; Novoselov, K.S. Detection of individual gas molecules adsorbed on graphene. *Nat. Mater.* **2007**, *6*, 652–655.

147. Gunlycke, D.; Lawler, H.M.; White, C.T. Room-temperature ballistic transport in narrow graphene strips.*Phys. Rev. B* **2007**, *75*, 085418:1–085418:5.

148. Ferrari, A.C. Raman spectroscopy of graphene and graphite: Disorder, electron–phonon coupling, doping and nonadiabatic effects. *Solid State Commun.* **2007**, *143*, 47–57.

149. Teague, M.L.; Lai, A.P.; Velasco, J.; Hughes, C.R.; Beyer, A.D.; Bockrath, M.W.; Lau, C.N.; Yeh, N.-C. Evidence for strain-induced local conductance modulations in single-layer graphene on SiO_2. *Nano Lett.***2009**, *9*, 2542–2546.

150. Guinea, F.; Horovitz, B.; Le Doussal, P. Gauge fields, ripples and wrinkles in graphene layers. *Solid State Commun.* **2009**, *148*, 1140–1143.

151. Han, M.Y.; Özyilmaz, B.; Zhang, Y.; Kim, P. Energy band-gap engineering of graphene nanoribbons. *Phys. Rev. Lett.* **2007**, *98*, 206805:1–206805:4.

152. Avouris, P.; Chen, Z.; Perebeinos, V. Carbon-based electronics. *Nat. Nanotechnol.* **2007**, *2*, 605–615.

153. Nakada, K.; Fujita, M.; Dresselhaus, G.; Dresselhaus, M.S. Edge state in graphene ribbons: Nanometer size effect and edge shape dependence. *Phys. Rev. B* **1996**, *54*, 17954–17961.

154. Son, Y.-W.; Cohen, M.L.; Louie, S.G. Energy gaps in graphene nanoribbons. *Phys. Rev. Lett.* **2006**, *97*, 216803–216804.

155. Kharche, N.; Zhou, Y.; O'Brien, K.P.; Kar, S.; Nayak, S.K. Effect of layer stacking on the electronic structure of graphene nanoribbons. *ACS Nano* **2011**, *5*, 6096–6101.

156. Chen, Z.; Lin, Y.-M.; Rooks, M.J.; Avouris, P. Graphene nano-ribbon electronics. *Phys. ELowDimens. Syst. Nanostruct.* **2007**, *40*, 228–232.

157. Stankovich, S.; Piner, R.D.; Nguyen, S.T.; Ruoff, R.S. Synthesis and exfoliation of isocyanate-treated graphene oxide nanoplatelets. *Carbon* **2006**, *44*, 3342–3347.

158. Schniepp, H.C.; Li, J.-L.; McAllister, M.J.; Sai, H.; Herrera-Alonso, M.; Adamson, D.H.; Prud'homme, R.K.; Car, R.; Saville, D.A.; Aksay, I.A. Functionalized single graphene sheets derived from splitting graphite oxide. *J. Phys. Chem. B* **2006**, *110*, 8535–8539.

159. Cai, J.; Ruffleux, P.; Jaafar, R.; Bieri, M.; Braun, T.; Blankenburg, S.; Muoth, M.; Seitsonen, A.P.; Saleh, M.; Feng, X.; *et al.* Atomically precise bottom-up fabrication of graphene nanoribbons. *Nature* **2010**, *466*, 470–473.

160. Li, X.; Wang, X.; Zhang, L.; Lee, S.; Dai, H. Chemically derived, ultrasmooth graphene nanoribbon semiconductors. *Science* **2008**, *319*, 1229–1232.

161. Dresselhaus, M.S.; Dresselhaus, G. Intercalation compounds of graphite. *Adv. Phys.* **2002**, *51*, 1–186.

162. Jiao, L.; Wang, X.; Diankov, G.; Wang, H.; Dai, H. Facile synthesis of high-quality graphene nanoribbons.*Nat. Nanotechnol.* **2010**, *5*, 321–325.

163. Naeemi, A.; Meindl, J.D. Performance Benchmarking for Graphene Nanoribbon, Carbon Nanotube, and Cu Interconnects. In Proceedings of International Interconnect Technology Conference, Burlingame, CA USA, 14 June 2008; pp. 183–185.

164. Balandin, A.A.; Ghosh, S.; Bao, W.; Calizo, I.; Teweldebrhan, D.; Miao, F.; Lau, C.N. Superior thermal conductivity of single-layer graphene. *Nano Lett.* **2008**, *8*, 902–907.

165. Ouyang, Y.; Wang, X.; Dai, H.; Guo, J. Carrier scattering in graphene nanoribbon field-effect transistors.*Appl. Phys. Lett.* **2008**, *92*, 243124:1–243124:3.

166. Wang, H.M.; Wu, Y.H.; Ni, Z.H.; Shen, Z.X. Electronic transport and layer engineering in multilayer graphene structures. *Appl. Phys. Lett.* **2008**, *92*, 053504:1–053504:3.

167. Lu, Y.H.; Chen, W.; Feng, Y.P.; He, P.M. Tuning the electronic structure of graphene by an organic molecule. *J. Phys. Chem. B* **2009**, *113*, 2–5.

168. Wang, X.; Li, X.; Zhang, L.; Yoon, Y.; Weber, P.K.; Wang, H.; Guo, J.; Dai, H. N-doping of graphene through electrothermal reactions with ammonia. *Science* **2009**, *324*, 768–771.

169. Wei, D.; Liu, Y.; Wang, Y.; Zhang, H.; Huang, L.; Yu, G. Synthesis of N-doped graphene by chemical vapor deposition and its electrical properties. *Nano Lett.* **2009**, *9*, 1752–1758.

170. Wehling, T.O.; Novoselov, K.S.; Morozov, S.V.; Vdovin, E.E.; Katsnelson, M.I.; Geim, A.K.; Lichtenstein, A.I. Molecular doping of graphene. *Nano Lett.* **2008**, *8*, 173–177.

171. Lee, E.J.H.; Balasubramanian, K.; Weitz, R.T.; Burghard, M.; Kern, K. Contact and edge effects in graphene devices. *Nat. Nanotechnol.* **2008**, *3*, 486–490.

172. Faugeras, C.; Nerriere, A.; Potemski, M.; Mahmood, A.; Dujardin, E.; Berger, C.; de Heer, W.A. Few-layer graphene on SiC, pyrolitic graphite, and graphene: A Raman scattering study. *Appl. Phys. Lett.* **2008**, *92*, 011914:1–011914:4.

173. Cervantes-Sodi, F.; Csanyi, G.; Piscanec, S.; Ferrari, A.C. Edge-functionalized and substitutionally doped graphene nanoribbons: Electronic and spin properties. *Phys. Rev. B* **2008**, *77*, 165427:1–165427:13.

Chapter 3

Nano-Scale Modelling and Simulation of Metal Wiredrawing by Using Molecular Dynamics Method

Ken-ichi Saitoh[1*], Youhei Sameshima[2], Syuhei Daira[2,3]

[1]*Department of Mechanical Engineering, Kansai University, Suita, Japan*

[2]*Graduate School of Science and Engineering, Kansai University, Suita, Japan*

[3]*Denso Techno, Co. Ltd., Obu, Japan*

ABSTRACT

In this paper, molecular dynamics (MD) simulations of nano-sized wiredrawing are performed. The wiredrawing is a traditional plastic working method, but there has not been any insight to develop it in a nano-sized scale. Therefore, to materialize the concept of the nano-sized wiredrawing, a numerical modelling is pursued at first in this paper, and the interatomic potential, a crystalline orientation, the drawing condition realized by a die geometry are thoroughly investigated. In particular, to reduce the friction between a wire and a die, a simple friction model for the MD analysis is newly proposed, where the interatomic interaction is adequately modified by a single factor ω. Then, the fruitful results are obtained by using $\omega = 0.1$. We checked the availability of such nano-sized MD simulation by constructing a two-dimensional wiredrawing model, at first. The analysis of atomic stress during drawing is also assessed. It is useful to use invariant of the atomic stress tensor, such as hydrostatic stress (average stress, σ_m) or von Mises equivalent stress (σ_{eq}). The former is related to the phase transformation from the body-centered-cubic (bcc) structure to the face-centered-cubic (fcc) one, which is found in the present MD simulation. It is observed that an initial α-iron crystal with bcc structure changes partially into the fcc phase. It is recognized that the phase transformation is caused by the positive hydrostatic stress values, which is occurring especially inside the die region. We observed that a lot of dislocation core structures occur in wiredrawing process and their existence and evolution are well related to the equivalent stress values.

Keywords: Molecular Dynamics, Iron and Steel, Wiredrawing, Plastic Working, Nanotechnology, Materials Processing, Crystalline Structure

1. INTRODUCTION

There are many plastic working processes that are important for industry. Above all, drawing (wiredrawing) process is extensively used to fabricate axisymmetric metallic bars and wires into a prescribed diameter. It is interesting that the wiredrawing is basically using only a die. So far, for example, drawn wires made up of iron and steel have been applied to many industrial products, such as tire codes, piano (music) wire, and suspension wire of huge bridge. Today, materials are trending toward very narrow and small size. For the purpose of utilizing wire structures in the field of MEMS (Micro Electro Mechanical System) or small-sized electronic/computer devices, the dimension of wire material should be as small as possible. From now on, a diameter required in such purpose will become less than one micrometer, i.e. less than several hundreds of nano-meters. Certainly, there is another cutting-edge method to synthesize very tiny wire using nano-structured template [1] . But, it has to be based on chemical reaction process and spontaneous crystal growth technique, so it is not so easy to control the fabricating process. Therefore, as an effective method for the production of very narrow and small wires, the traditional plastic working technique using the die may be applied, by downsizing the wiredrawing process into a nano-meter scale. Indeed, it is, of course, just an idea with possibility now, but in near future it will be categorized as a prospective extension of the traditional wiredrawing method. So far, the smallest steel wire made by commercial drawing has just reached to several micrometers in diameter. Accordingly, "nanoscale wiredrawing" is one of prospective technologies. However, now, we do not have any evidence for feasibility of nano-meter wire drawing technique. For this reason, we need first to study the possibility of this new technique, and we should predict the material behaviors by numerical simulation.

In this study, the molecular dynamics (MD) method, which is one of theoretical and computational techniques for materials modelling and simulation, is applied to the nano-sized wiredrawing process. In numerical modelling, except for MD method, usually, finite element (FE) analysis, which is definitely based on some mathematical elastic-plastic continuum theory, has been widely applied to investigate the wiredrawing process. In recent years, since information about the crystalline change of metallic materials during wiredrawing is very crucial, the FE analysis including crystal plasticity theory (CPFE) [2] [3] is well being applied to the study. Though the FE analysis can produce many practical results, its material model must be constructed in mathematical fashion and the interaction between the die and the wire is still inevitably ambiguous. Namely, microscopic physics of a wire material is not sufficiently modeled by the conventional FE or even precise CPFE framework. The conventional FE is also unsuitable to analyze the texture and defects which are being nucleated and developing in the wire. To the contrary, the MD method is capable of modelling a wire material from atomic-scale physics and so the method has an advantage in predicting microscopic phenomena. It is well known that there are some drawbacks of MD, however. At the present, the largest three-dimensional computational size of atomic system treatable in the MD method is just limited to the range around several hundreds nano-meters. However, as stated above, there is a trend to make smaller metallic wire and this fact requires us to simulate the phenomena from the atomic scale to obtain a clear insight about wiredrawing. For those reasons, the present authors suppose that it is very worthwhile using the MD method to analyze the wiredrawing process.

So far, plasticity or plastic working process has been investigated by the MD method as follows. As for analyses of plastic working application, for example, the rolling [4] and the cutting [5] of metallic materials have been studied. As for unknown phenomena concerning plasticity, to name a few, dynamic recrystallization in Cu-Al system [6] , pencil glide dislocation motion in bcc (body-centered cubic) crystal [7] , and martensitic phase transformation of alloy [8] , have been studied. Unfortunately yet, there has not been any MD study concerning wiredrawing. In the wiredrawing process, the force and the contact conditions subjected to the wires are tremendously strong. Therefore, the wiredrawing is one of strong plastic working techniques, where the material is subjected to very strong multi-axial stress and strain. A similar viewpoint can be found in the MD analysis concerning severe plastic deformation (SPD) processes [9], many of which have been experimentally invented as novel plastic processing method. Also, the nano-sized imprint (nano-imprint) technique of copper has been attempted in the MD simulation [10] , while the material implemented there is generally a soft polymer or a resin. Once again, to our knowledge, there has not yet been any MD study of nano-sized wiredrawing.

In this study, iron and steel wires are focused on, for they are extensively used with the wiredrawing in industry. In iron and steel, an impurity addition such as carbon, etc., is usually carried out. Consequently, both strength and ductility are enhanced at the same time, thanks to an eutectic phase called a pearlite phase. The pearlite phase is composed of ferrite (α-Fe) and cementite (Fe_3C) crystal as sub-structures inside. In the present MD simulation, however, some drastic simplification in the computational model is inevitable, especially for the modelling of interatomic potential function. So, at this point, it is reasonable to use a single crystal model comprising just the ferrite crystal and then a reliable potential function just between iron atoms will be successfully adopted.

In this study, an atomic-scale wiredrawing model is constructed for the MD simulation. The new knowledge of atomic mechanism in the wiredrawing which is presented in this study is useful for the future micro-scale or nano-scale wiredrawing process. Ideally, the wiredrawing includes an axisymmetric condition and so a full three-dimensional model would be optimum. But, for the simplification of the MD model, a two-dimensional drawing model using periodic boundary condition is adequate now and here. The realistic crystal rotation and the motion of lattice defects will be precisely captured by upcoming study of a full three-dimensional model [11].

As one of topics in our study of MD modelling, a friction model is proposed and is evaluated. As frequently observed in MD analyses, a friction force between nano-sized solids becomes tremendously large, because of larger relative contacting area between surfaces. In our model, the interatomic potential between the die and the wire atoms is modified to conduct the drawing smoothly with a moderate lubricating condition. Since the present friction model requires somewhat parametric study, the first evaluation is provided here.

Understanding the stress or strain distribution inside a drawn wire, including the residual stress distribution, is important in discussing the result of wiredrawing process. In the history of studies of the wiredrawing [12] -[16] , hydrostatic and (von Mises) equivalent stress components are key factors which should be focused on. This paper shows that the atomic motion and behavior of lattice defects are closely related to their stress distribution. We also affirm the result of crystalline phase transition between bcc and fcc (face-centered cubic) in microscopic scale. This behavior has not yet observed experimentally in the drawing of steel wires. The present study shows that the mechanism of phase transition is also closely related to the stress distribution inside the wire.

2. THEORY AND METHODS

2.1. Molecular Dynamics

Molecular dynamics (MD) is a microscopic numerical simulation technique, where the time evolution of atomic or molecular system is obtained. The interatomic interaction between atoms or molecules is prescribed. The equations of motions are numerically integrated, with a certain time increment (usually in the order of femto- second), to update positions and velocities of individual atoms. Using the atomic trajectories in the system, the development of crystal structure and material phase is captured. Accordingly, dynamics of phase transition can be simulated directly, and transport coefficients such as diffusion constant can be estimated effectively.

The equations of motion are given by Equation (1),

$$m_i\left(\mathrm{d}^2\boldsymbol{r}_i/\mathrm{d}t^2\right)=\boldsymbol{F}_i=-\partial E_{\mathrm{tot}}/\partial\boldsymbol{r}_i\,, \tag{1}$$

where m_i, r_i, F_i, and t are atomic mass, position, interatomic force, and time, respectively. The MD simulation is sometimes called a deterministic method, which is different from any stochastic method such as Monte Carlo method. The interatomic force is straightforwardly determined as the derivative of the potential energy of the total system E_{tot}. Though the MD method provides a high-resolution view of the material, the calculated size is only limited to very tiny portion of the material. In addition, the interatomic potential function is just an approximation of real material system. However, those drawbacks are being somewhat resolved by introducing ab- initio (first principle) method involving quantum mechanical formulation and also by utilizing a parallel algorithm to enable very larger-scale computations.

2.2. Interatomic Potential Function of Iron and Steel

The crystal structure of iron is bcc in the standard temperature and pressure, while it changes to fcc in high temperature or stress application. Accordingly, most of potential functions are formulated in the bcc structure. The potential function of the embedded atom method (EAM) is adopted in the present study. The EAM can reproduce the many-body effect around a metallic atom. We use the Finnis-Sinclair type potential (FS potential) [17] , which has been extensively used in other MD studies, because of its high accuracy in system energy and atomic behavior of iron system. In the FS potential formulation, the total energy of the system is given by,

$$E_{\mathrm{tot}} = E_{\mathrm{m}} + E_{\mathrm{p}} = \sum_{i=1}^{N} AF\left(\rho_i\right) + \left(1/2\right)\sum_{i=1}^{N}\sum_{j\neq i}^{N}\phi\left(r_{ij}\right), \tag{2}$$

where E_m means the many-body term, ρ_i is an electron density made up of interactions from surrounding atoms, and $F(\rho)$ means a functional of the electron density. The second term E_p is a summation of usual pairwise interactions $f(r)$.

2.3. The Concept of Nano-Scale MD Wiredrawing Model

Generally speaking, the wiredrawing is one of traditional plastic working processes, in which bar or rod is forced to go through the die, and is eventually given a narrower diameter (i.e. it is shrunk in diameter and is stretched in the longitudinal direction). Consequently, the wire material is given higher strength and ductility as well as a fine surface finish. The purpose of the present study is to build a nano-meter-sized simulation model for the wiredrawing process. The idea that a hollow die is used in the wiredrawing is straightforwardly adopted. Therefore, whether it will be actually better choice or not is, in fact, left to the future experimental study.

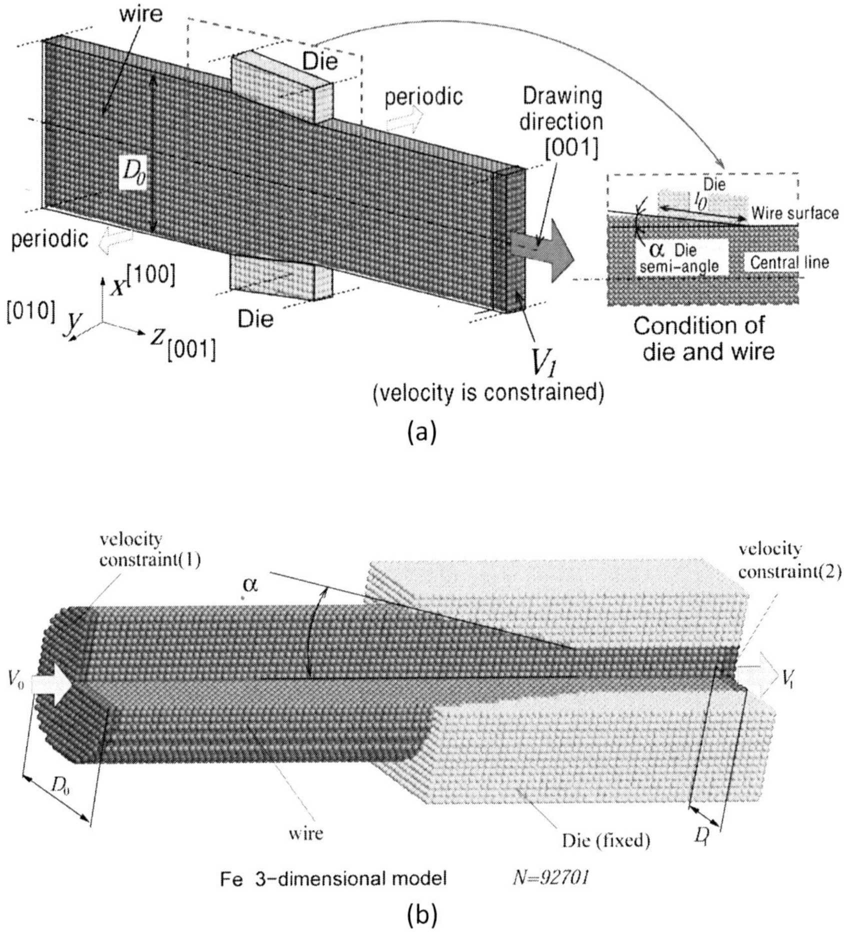

Figure 1. Molecular dynamics models for wiredrawing. (a) An example of the present 2-d model(a model of the drawing in [100] direction); (b) Another example of 3-d model [11] [18] (this is not calculated here).

2.4. Nano-Scale MD Wiredrawing Model

It is true that a three-dimensional model is more precise for the wiredrawing as shown in Figure 1(b) which is being used in another study [11] [18] . But, in three-dimension, understanding the behavior of crystal transition and defects is quite complicated yet. Therefore, in the present study, an approximated two-dimensional model of the wiredrawing, as shown in Figure 1(a), is constructed and is applied to the MD simulation. The MD calculation is carried out in three-dimensional space, but the periodic boundary condition is imposed on in one direction (in this case, y direction) which is perpendicular to the drawing direction (in this case, z direction). This model seems like the rolling process, but, since the deformation in periodic direction (y direction) is constrained, the stress state tends to be tri-axial similar to the wiredrawing.

The die semi-angle is 5.0 degrees and the length of the die slope is 5.0 nm, which are fixed through the present study. In fact, the die semi-angle in between 5 and 7 degrees is proved to be optimum in axisymmetric drawing by analytic consideration of theoretical plasticity. As recognized well, the choice of crystalline orientation relative to the drawing direction affects the mechanical property of the drawn wire. Therefore, [100] [111] orientations are compared here. All the calculation conditions are summarized in Table 1.

2.5. Construction of Friction Model

The plastic working technique usually copes with the problem of the friction between the die and the wire material [19] . Since the relative contact area is larger in nano-sized scale, the effect of friction becomes larger. It is supposed that the future nanotechnology should overcome the friction problem as a primal subject. The developing study of nano-imprint technique is one of the challenges [10].

In this study, it is proposed that the friction is reduced by modifying the potential function between the die and the wire atoms as shown in Figure 2. The interatomic interaction is somewhat needed to make the friction, but it is reduced not to adhere each other. The general formulation is shown as follows. For example for pairwise potentials, the cohesive energy (interatomic binding energy) is determined by the intensity of the potential function itself. Therefore, the potential function is reduced by an arbitrary factor ω, which is less than unity (ω <

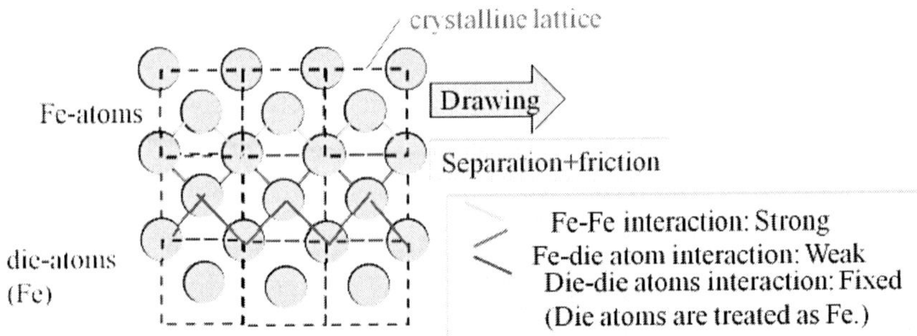

Figure 2. The concept of friction model of MD in wiredrawing process (Interaction as to die atoms are that of Fe atoms).

Table 1. Calculation conditions and materials properties.

Property	Unit	Value
The model of drawing in [100] direction		
The number of total atoms		91,885
The number of fixed atoms (wire)		1362
The number of fixed atoms (dies)		15,968
The model of drawing in [110] direction		
The number of total atoms		91,919
The number of fixed atoms (wire)		973
The number of fixed atoms (dies)		16,132
The model of drawing in [111] direction		
The number of total atoms		91,911
The number of fixed atoms (wire)		973
The number of fixed atoms (dies)		16,089
Size $x \times y \times z$	nm	9.8796 × 9.8796 × 27.5184
Lattice constant a_0 (α-Fe)	nm	0.28665
Length of die land l_0	nm	5.00
Diameter D_0	nm	6.8796 (=24 a_0)
Die semi-angle α	deg.	5.00
Outlet velocity V_1	m/s	50.0 (fast)/10.0 (slow)
Time increment Δt	fs	2.0
Temperature T	K	10.0

1). At this point, the factor ω is completely ad-hoc parameter and it should be fitted by ab-initio calculation or experimental result in the future. So far, we have evaluated the adequate value of ω in our MD simulation. As an consequence, it is found that ω should be less than 0.1, and then it shows good results for most of wiredrawing conditions.

Generally, the original potential function V(r) is modified by Equation (3).

$$V_a(r) = \omega \cdot V(r) \tag{3}$$

The many-body potential is composed of the many-body term and the pairwise term, as shown in Equation (2). Interatomic bonding is largely contributed by the many-body term and it is usually a monotonically increasing function of the electron density ρ(r). Thus, for the FS potential in this study, ρ(r) is directly multiplied by the factor ω introduced above. The expression of the FS potential including the friction condition is given by Equation (4) together with Equation (2),

$$\rho_{i,a} = \begin{Bmatrix} \omega \cdot \rho(r) & (\text{wire-die}) \\ \rho(r) & (\text{wire-wire}) \end{Bmatrix} = \sum_{j=1}^{N(\text{wire-die})} \omega \cdot \psi(r_{ij}) + \sum_{j=1}^{N(\text{wire-wire})} \psi(r_{ij}), \tag{4}$$

where ψ(r) is a pairwise function for the electron density. The $\omega < 0.1$ is found to show a relevant friction in the many-body case as well.

2.6. Analyzing Method of the MD Output

CNA Method: In order to detect a crystal type and its orientation, the common neighbor analysis (CNA) can be used effectively [20] [21] . By using the CNA, one can identify each atom as fcc, bcc or the other crystalline structures. Besides, by the CNA, we can determines atoms near a lattice defect such as dislocation, grain boundary or surface.

Atomic Stress: It is possible to calculate the atomic stress tensor (atomic stress) in MD calculation. The atomic stress tensor is derived along with the continuum concept of stress tensor. When an individual atomic volume is assumed, then strain energy stored inside the volume is evaluated from interatomic potential. The energy is mathematically differentiated by the strain tensor which assumes homogeneous deformation and is expressed by the distance between pair of atoms. This means that atomic stress corresponds to increase of the elastic strain energy per unit volume. Thus, the general formulation of atomic stress of atom i is defined by Equation (5).

$$\sigma_i^{\alpha\beta} = \{1/(2\Omega)\} \sum_{j=1}^{N} \left(\partial E_i(r)/\partial r\right)\Big|_{r=r_{ij}} \cdot r_{ij}^{\alpha} r_{ij}^{\beta} / r_{ij} , \tag{5}$$

where $r_{ij}^{\alpha} = r_i^{\alpha} - r_j^{\alpha}$ is the difference of position vectors between i and j atoms, r_{ij} is their distance. $E_i(r)$ is an effective energy contribution from i atom. Ω is the atomic volume for the reference structure. Generally speaking, the summation of many-body energy is unable to be divided into the energy of each atoms. However, for FS potential, we can do it only in considering an effective potential,

$$E_i(r_{ij}) = F(\rho_i) + \sum_{j=1, j\neq i}^{N} \phi(r_{ij}) \tag{6}$$

In finite temperature, atomic stress inevitably contains large fluctuation due to the small number of atoms. So, it needs to be averaged in volume or in time interval to give continuous distribution. The averaged stresses are displayed without the atomic subscript i, like $\sigma^{\alpha\beta}$, and we use these values throughout this paper and call them only "stress".

Once the components of the stress with regard to a set of coordinate axes (x, y, z) are obtained, effective stress invariants are calculated from them. They are, for instance, hydrostatic stress (mean stress) σ_m and von Mises equivalent stress σ_{eq}. σ_m is also the first invariant of stress tensor, defined by,

$$\sigma_{\mathrm{m}} = \left(\sigma^{xx} + \sigma^{yy} + \sigma^{zz}\right)/3 = -p , \tag{7}$$

where p means hydrostatic pressure. The σ_m represents a tri-axial condition of the stress and it is related to the volume change. For example in the wiredrawing process, with increase of the σ_m in positive regime (i.e. tensile stress, $\sigma_m > 0$), an internal fracture (also called central burst or "cuppy" failure in the actual process) tends to take place. On the other hand, the σ_{eq} is related to the second invariant of stress. It combines the deviatric stress components and is defined by,

$$\sigma_{eq} = \sqrt{\left\{\left(\sigma^{xx} - \sigma^{yy}\right)^2 + \left(\sigma^{yy} - \sigma^{zz}\right)^2 + \left(\sigma^{zz} - \sigma^{xx}\right)^2\right\}/2 + 3\left\{\left(\sigma^{xy}\right)^2 + \left(\sigma^{yz}\right)^2 + \left(\sigma^{zx}\right)^2\right\}}, \tag{8}$$

Physically, the σ_{eq} value reflects a nucleation event of dislocation and its motion in a crystal. In the atomic system in MD simulation, the σ_{eq}, which is evaluated locally in an atomic system, exhibits the intensity of driving force for the nucleation or the mobility of dislocations.

3. RESULTS AND DISCUSSION

3.1. Effect of the Factor ω on the Friction Behavior

Wiredrawing simulations are performed with the variety of friction factors, ω = 1.0, 0.1, 0.01 and 0.001, and the dependency on friction factor ω is investigated. The maximum stress value for components σ^{zz}, σ_m, and σ_{eq}, which are averaged over all atoms in the wire, are shown in Figure 3(a). Moreover, Figure 3(b) shows the time-averaged stress components. It is clear that the larger ω is, the larger the stresses are. It means that resistance of the wire material during the drawing process, is totally raised by larger ω near unity. The resistance is possibly caused by the friction at the wire surface between the die and the wire atoms as well as the plastic deformation occurring inside the wire. If we assume that the resistance due to plasticity provides a constant resistance, the friction at the wire surface is responsible for the change in the total resistance in the drawing process. Besides, the results for ω = 0.01 and ω = 0.001 are not so different and their difference in stresses is almost zero. Thus, we can apply the friction model with the factor ω = 0.001 in all the MD simulations presented here.

3.2. Stress Distribution and Its Change in Wiredrawing Process

3.2.1. Drawing Stress

Figure 4 shows the value of σ^{zz} which is the normal stress component in the direction of wiredrawing (i.e. σ^{zz} is a drawing stress) and σ^{xx} which is in the perpendicular direction and not in the direction of periodicity of the model. After a start-up period when the length 1.5 nm of the wire has been drawn, the stress component σ^{xx} tends to be negative (compressive) and then becomes stationary. On the other hand, the drawing stress, σ^{zz}, tends to be positive (tensile) and becomes stationary. The value of σ^{zz} reaches 6.0 GPa at maximum, the magnitude of which is about twice the maximum value of the magnitude of σ^{xx}. As for comparison among crystalline orientations, the wiredrawing in [100] direction shows the largest stress values among three. If we assume the same effect of the friction at the surface, this fact means that the crystal arranged in [100] orientation shows the largest resistance to the plastic deformation among three orientations.

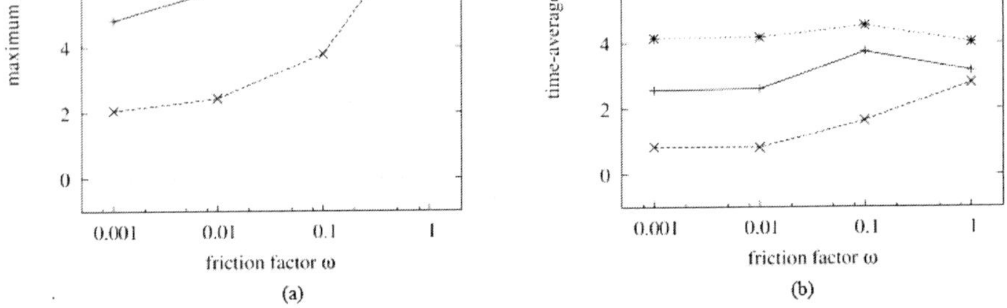

Figure 3. Relation between friction factor ω and (a) maximum stress; (b) time-averaged stress in wiredrawing.

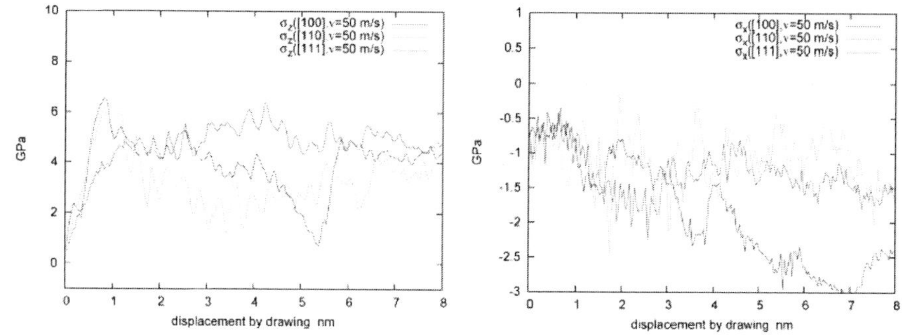

(a) (b)

Figure 4. Time transition of stresses occurring in wire (drawing direction σ^{zz}/perpendicular direction σ^{xx} (one of not periodic directions), case: v = 50 m/s). (a) σ^{zz}; (b) σ^{xx}.

3.2.2. Hydrostatic Stress

In order to comprehend characteristics of the stress state in the wire, it is highly reasonable to look into both hydrostatic stress, σ_m, and equivalent stress, σ_{eq}. The normal stress components, σ^{xx}, σ^{yy} or σ^{zz} may be either tensile or compressive, but an averaged tri-axial stress state, i.e. $\sigma_m$tends to be in tension or in compression. The hydrostatic stress, σ_m, corresponds only to an elastic deformation, whereas equivalent stress, σ_{eq}, is derived from the deviatric stress tensors which are a residue of stress tensor in excluding isotropic contribution. Accordingly, the atomistic value of σ_{eq}shows a measure of the plastic working which those atoms have been experienced during the wiredrawing.

Time transition of σ_m and σ_{eq} during drawing is visualized in Figure 5. It is understood that σ_m is always positive (tensile stress). However, when the drawing speed is slower (condition (B)), the magnitude of σ_m is converging into zero. This means that, in the case with a higher drawing speed (such as v = 50 m/s, Figure 5(a)), the deformation velocity in plasticity is defeated by the traveling speed of the wire. As a result, the plastic deformation has not fully been carried out even after the wire comes out of the die region. To the contrary, in a slower drawing case, there is a plenty of time for the atoms to undertake a plastic deformation, including a nucleation of lattice defect such as dislocation. The whole crystalline structure is relaxed and the hydrostatic stress σ_m drops soon after that, in this case. As shown in Figure 5(b), the plastic deformation can proceed, not increasing the stress σ_{eq} and keeping the magnitude of σ_m at a low level.

Figure 6 shows the stress distribution when it is viewed in the cross-section of the wire (projected onto xz plane). This is the case of wiredrawing with [100]-orientation and v = 50 m/s. Those pictures are arranged as follows: (A) normal stress in drawing direction σ^{zz}, (B) hydrostatic (mean) stress σ_m, and (C) equivalent stress σ_{eq}, from the top to the bottom. Also, the pictures at 40, 80 and 120 ps are arranged from the left to the right.

The drawing stress σ^{zz} possesses a positive value after passing through the die region, whereas the trailing part shows a negative value (compressive). These plus and minus in the stress value seems balanced in the total region of the wire. The mean stress σ_m shows a larger value, especially near the exit of the die region, and it is decreasing with increase of the distance from the exit of the die. Besides, a v-shaped region of the compressive stress exists at the entrance of the die region. Since the wire is always compressed by the inclined surface of the die, it is reasonable that the atoms in the wire have always compressive stress.

Equivalent stress σ_{eq} shows a large value at the center of the die region. The stress value of the wire is increasing with increase of the distance from the exit of the die. It seems natural that the σ_{eq} is large around the region in which the wire material is experiencing a plastic deformation. As an example, an isolated dislocation structure is observed at the lower part in the cross-sectional picture which is taken outside the die region, as shown in the figure at 120 ps. Discussion concerning the structure and the dynamics of dislocations will be shown in the subsequent section in relation to the CNA analysis.

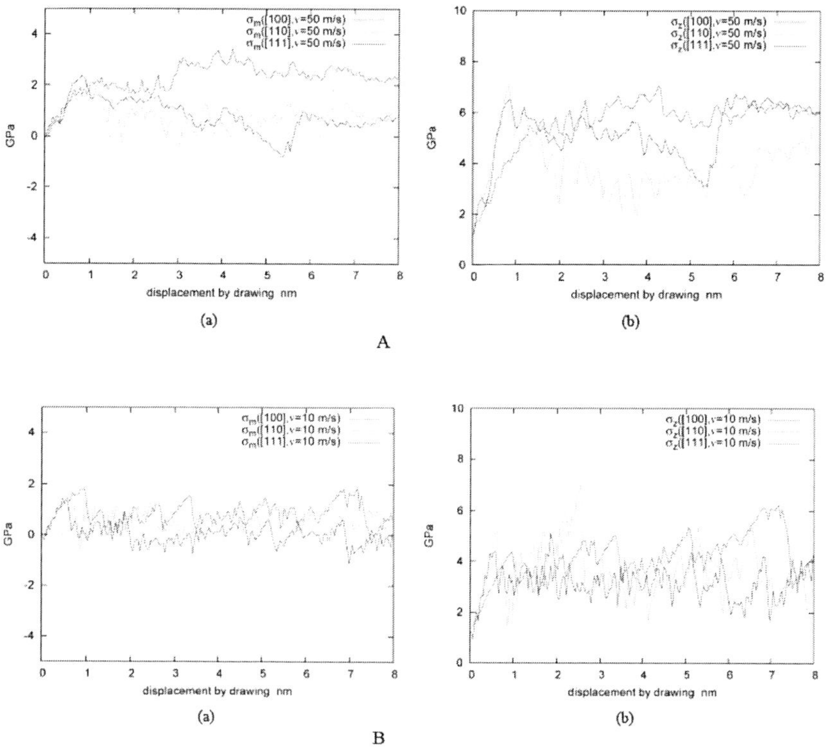

Figure 5. Time transition of stresses occurring in wire (hydrostatic component σ_m and equivalent stress component σ_{eq}). A) Fast drawing velocity v = 50 m/s, (a) hydrostatic stress σ_m; (b) equivalent stress σ_{eq}. B) Slow drawing velocity v = 10 m/s, (a) hydrostatic stress σ_m; (b) equivalent stress σ_{eq}.

3.3. Results of Crystal Change by Common Neighbor Analysis (CNA)

Figure 7 shows the distribution and its transition of the crystalline state which is analyzed by common neighbor analysis (CNA) during wiredrawing. It is found that, when the drawing velocity is high, there occurs a phase transformation from bcc into fcc in some part of the wire. This fact will be discussed in more detail later in this paper. For a slow drawing velocity, this type of phase transition certainly takes place, but the amount of it is much smaller compared with the fast drawing. Anyway, for both velocities, however, the transformed fcc phase does not stay alive so long time. Most of the transformed fcc phase disappears and changes back into a bcc phase spontaneously. This return event of the crystalline phase is observed, after the transformed fcc region of the wire just exits from the die region. It is guessed that the driving force needed to the phase transformation is no longer applied outside the die region.

Figure 8 shows the time transition of the ratio of fcc atoms to the whole wire atoms. As can be seen in the figures, the phase transformation takes place instantly after the drawing starts. There is a tendency that a higher drawing velocity promotes more atoms which are relating to the phase transformation.

Figure 9 shows a selected region of the atomic configurations including a dislocation core, which are analyzed by the CNA method. The arbitrarily selected time is 120 ps after the wiredrawing started.

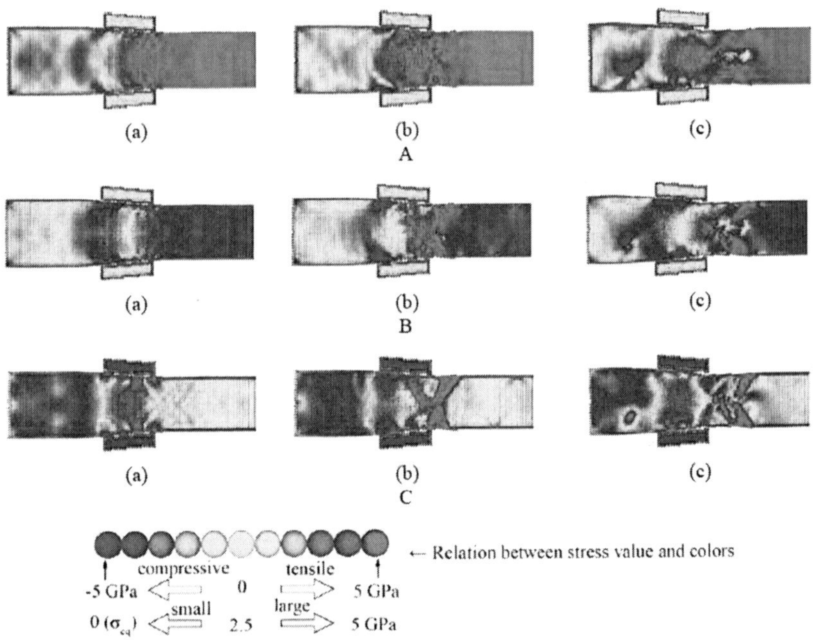

Figure 6. Stress distribution (σ^{zz}, σ_m, σ_{eq} in drawing process (v = 50 m/s, the drawing in [100] direction)). A) σ^{zz} drawing stress, (a) 40 ps, (b) 80 ps, (c) 120 ps; B) σ_m hydrostatic (mean) stress, (a) 40 ps, (b) 80 ps, (c) 120 ps; C) σ_{eq} von Mises equivalent stress, (a) 40 ps, (b) 80 ps, (c) 120 ps.

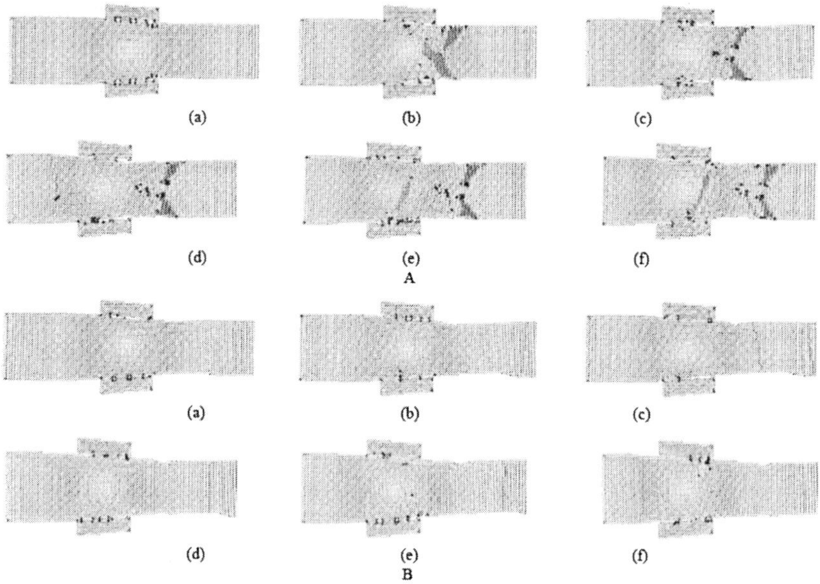

Figure 7. Distribution and its transition of atomic crystalline state: results of CNA analysis during the wiredrawing (v = 50 m/s and 10 m/s, the drawing in [100] direction). The colors are being used as follows: yellow = bcc atom, gray = fcc atom, blue = low-coordination atom (surface or amorphous), and brown = high-coordination (defected) atom). A) v = 50 m/s (fast), (a) 40 ps, (b) 80 ps, (c) 100 ps, (d) 120 ps, (e) 140 ps, (f) 160 ps; B) v = 10 m/s (fast), (a) 40 ps, (b) 80 ps, (c) 100 ps, (d) 120 ps, (e) 140 ps, (f) 160 ps.

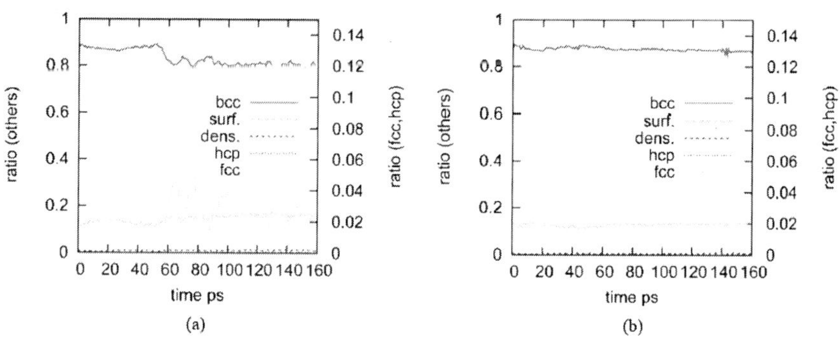

Figure 8. Time transition of fcc-atom ratio (the case of the drawing in [100] direction). (a) v = 50 m/s (fast); (b) v = 10 m/s (slow).

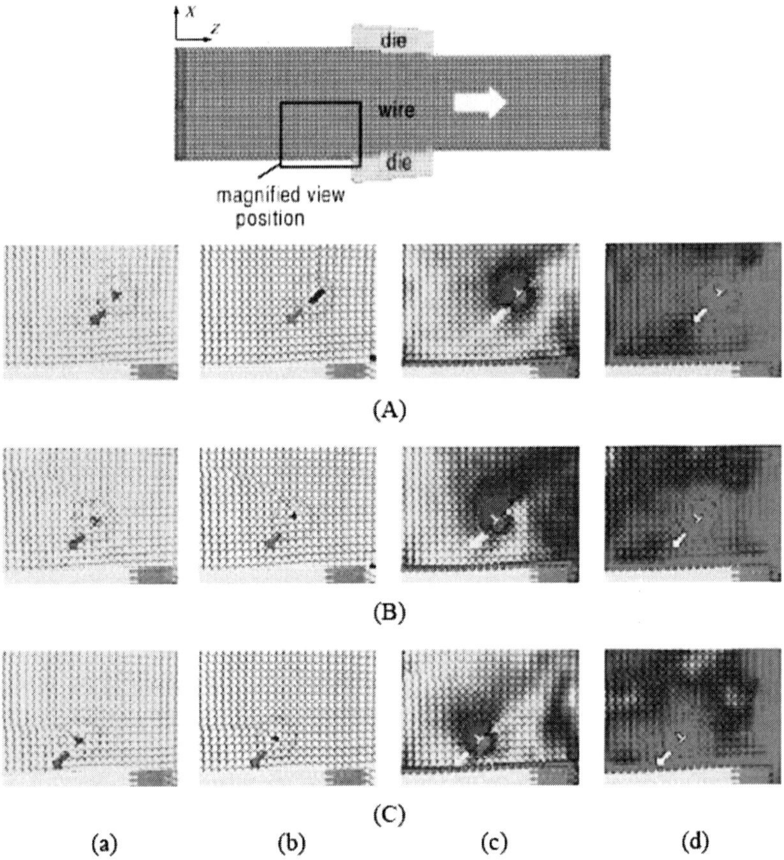

Figure 9. Comparison of visualization pictures around an isolated dislocation structure (at v = 50 m/s, in the case of the drawing in [100] direction). (A) at t = 119.20 ps; (B) at t = 120.00 ps; (C) at t = 120.80 ps. (a) atomic arrangement; (b) CNA analysis; (c) hydrostatic (σ_m); (d) equivalent stress (σ_{eq}).

This case is for the drawing of [100]-orientation. The series of pictures in Figure 9 are displayed to compare between the atomic configuration, CNA analysis, the distribution of hydrostatic stress, and the distribution of equivalent stress, respectively. It is understood, from the atomic arrangement and the stress distribution, that this dislocation contains edge and screw components at the same time. The magnitude of the stress value is especially larger in the dislocation core. The dislocation is moving from upper-right to lower-left in angle of 45 degrees with regard to the horizontal axis in these pictures. By viewing pictures of the CNA analysis, it is confirmed that this dislocation is an isolated one, and it is atomistically composed of both less-coordinated and excessive-coordinated atoms which are located above and beneath the slip plane, respectively.

The hydrostatic stress of the atoms located above and beneath the slip plane have the opposite sign each other. The equivalent stress exhibits a large value widely around the dislocation core. The surface of the wire is seen at the bottom of these figures. When the dislocation reaches at the surface, the lattice imperfection due to the dislocation structure resolves naturally into a perfect crystal, where the wire surface is contacting strongly to the die surface. These MD results of the wiredrawing show that the events concerning generation, movement and annihilation of dislocation are all performed in the nano-sized scale and in a very short time.

Figure 10 shows stress components (hydrostatic stress σ_m, equivalent stress σ_{eq} and drawing stress σ^{zz}), all of which are averaged over atoms in each CNA categories. In the condition shown here (the [100]-oriented drawing and the drawing velocity of v = 50 m/s), it is observed that the phase transition into the fcc crystal occurs on a large area in this cross-section. All the stress values indicated in Figure 10 are averaged and normalized by the total average. At that moment, the ratio of categories is as follows: bcc = 85%, fcc = 3%, hcp (hexagonal close- packed) = 0%, surface (less coordinated) = 12%, and grain boundaries and defects (excessively coordinated) ≤ 1%. The fcc phase occupies relatively a large volume, but the fcc atoms there do not transform into a hcp phase (the hcp phase means regional stacking fault of the fcc crystal) at all. The hydrostatic stress is generally larger in the fcc-transformed region. The equivalent stress is also generally larger in the disordered (defected) region, such as the phase boundary around the fcc-transformed region, dislocation, or grain boundary.

From these facts, it is concluded that the hydrostatic stress component is responsible for the bcc → fcc phase transition. It can be said that the iron atomic system in our MD model undergoes a process of the severe plastic deformation (SPD). It also understood that such a strong plasticity condition can be realized by a nano-sized wiredrawing condition.

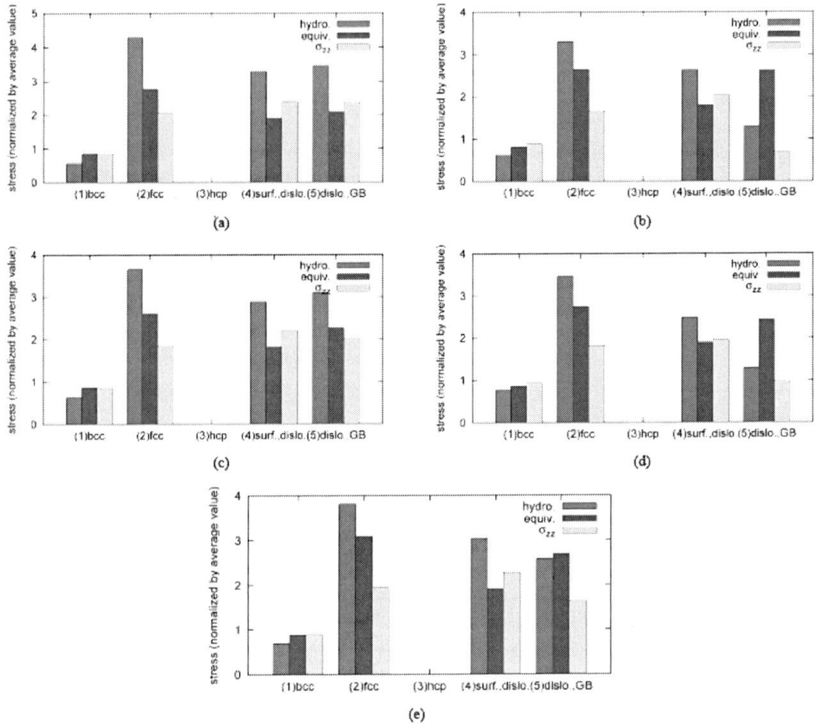

Figure 10. Relation between CNA categories and stress components (v = 50 m/s, the drawing in [100] direction). (a) 72.00 ps; (b) 76.00 ps; (c) 80.00 ps; (d) 84.00 ps; (e) 88.00 ps.

4. CONCLUSIONS

In order to assess the possibility of nano-sized wiredrawing, molecular dynamics (MD) modelling of α-iron is carried out. In the modelling, potential function, crystalline orientation and geometry of die etc. have to be considered so as to succeed in wiredrawing. In particular, introduction of friction model is inevitable. In this paper, to simplify the material behavior, the two-dimensional MD models with three different crystalline orientations are compared. Following results are obtained.

1) The interatomic function which is modified by the factor ω works well for reducing the friction between wire and die. The variety of ω values is compared. It is resulted that the smaller ω is, the smaller the resistance in drawing is obtained.

2) The invariant of atomic stress tensor is useful to recognize the status of atoms in the wire. Both hydrostatic stress σ_m and von Mises equivalent stress σ_{eq} are evaluated.

3) Phase transformation into fcc crystal occurs in wiredrawing of α-iron. It is caused by large hydrostatic stress σ_m in tensile regime (plus sign). The fcc phase is generally short-lived and is maintained only inside die region.

4) Dislocation structures are observed in wiredrawing. Atomistic equivalent stress clearly exhibits the core structure of a dislocation. A set of dislocation behaviors including nucleation, movement and annihilation at the wire surface completes in very small period.

ACKNOWLEDGEMENTS

This study is partly supported by the "Strategic Project to Support the Formation of Research Bases at Private Universities: Matching Fund Subsidy from MEXT (MInistry of Education, Culture, Sports, Science and Technology) (2012-2014)". The authors also acknowledge Nippon Steel & Sumitomo Metal Co. Ltd. and Kurimoto Ltd. for financial support.

REFERENCES

1. Tang, Z. and Kotov, N.A. (2005) One Dimensional Assemblies of Nanoparticles: Preparation, Properties, and Promise. Advanced Materials, 17, 951-962

2. Roters, F., Eisenlohr, P., Hantcherli, L., Tjahjanto, D.D., Bieler, T.R. and Raabe, D. (2010) Overview of Constitutive Laws, Kinematics, Homogenization and Multiscale Methods in Crystal Plasticity Finite-Element Modeling: Theory, Experiments, Applications. Acta Materialia, 58, 1152-1211.

3. Yoshida, T., Morita, Y. and Nakamachi, E. (2013) Crystal Texture Evolution Analyses in Drawing Process by Using Multi-Scale Finite Element Method. Proceedings of JSME (88th Kansai Branch Annual Meeting), 134-1, (5-)23.

4. Nakatani, A. (2005) Plastic Deformation Analysis of Nanostructured Metal Using Molecular Dynamics. Proceedings of JSME (2005 Annual Congress), 8, 470-471.

5. Chu, C.-Y. and Tan, C.-M. (2009) Deformation Analysis of Nanocutting Using Atomistic Model. International Journal of Solids and Structures, 46, 1807-1814.

6. Karthikeyan, S., Agrawal, A. and Rigney, D.A. (2009) Molecular Dynamics Simulations of Sliding in an Fe-Cu Tribopair System. Wear, 267, 1166-1176.

7. Ngan, A.H.W. and Wen, M. (2001) Dislocation Kink-Pair Energetics and Pencil Glide in Body-Centered-Cubic Crystals. Physical Review Letters, 87, Article ID: 075505.

8. Saitoh, K. and Liu, W.K. (2009) Molecular Dynamics Study of Surface Effect on Martensitic Cubic-to-Tetragonal Transformation in Ni-Al Alloy. Computational Materials Science, 46, 531-544.

9. Dan, T. and Saitoh, K. (2012) Microstructure Evolution in Polycrystalline Metal under Severe Plastic Deformation by Strain-Controlled Molecular Dynamics. Journal of Solid Mechanics and Materials Engineering, 6, 48-60.

10. Pei, Q.X., Lu, C., Liu, Z.S. and Lam, K.Y. (2007) Molecular Dynamics Study on the Nanoimprint of Copper. Journal of Physics D: Applied Physics, 40, 4928-4935.

11. Saitoh, K., Sameshima, Y., Takuma, M. and Takahashi, Y. (2014) Atomistic Simulation of Crystal Change and Carbon Diffusion in Nano-Sized Wiredrawing of Pearlitic Steel. Technische Mechanik, 32, in press.

12. Nakagiri, A., Yamano, T., Konaka, M., Yoshida, K. and Asakawa, M. (2000) Chevron Crack and Optimum Drawing Condition in the Diagram of Mean Stress and Die-Wire Contact Length Ratio by FEM Simulation. 2000 Conference Proceedings of the Wire Association International, Inc., 75-82.

13. Atienza, J.M., Martinez-Perez, M.L., Ruiz-Hervias, J., Mompean, F., Garcia-Hernandez, M. and Elices, M. (2005) Residual Stresses in Cold Drawn Ferritic Rods. Scripta Materialia, 52, 305-309.

14. McAllen, P.J. and Phelan, P. (2007) Numerical Analysis of Axisymmetric Wire Drawing by Means of a Coupled Damage Model. Journal of Materials Processing Technology, 183, 210-218.

15. Wright, R.N. (2009) Center Bursts—A Review of Criteria. Wire Journal International, 42, 80-84.

16. Yoshida, K. and Koyama, R. (2012) Reduction of Residual Stress of Drawn Wires. Wire Journal International, 45, 56- 60.

17. Finnis, M.W. and Sinclair, J.E. (1984) A Simple Empirical N-Body Potential for Transition Metals. Philosophical Magazinc A, 50, 45-55.

18. Saitoh, K., Daira, S. and Sameshima, Y. (2012) Nano-Scale Modelling and Simulation of Metal Wiredrawing by Using Molecular Dynamics Method. Proceedings of International Conference on Materials Processing and Technology 2012 (MAPT2012), Hawaii, 28-29 June 2012, 211-216.

19. Wright, R.N. (2010) Wire Technology "Process Engineering and Metallurgy". Butterworth-Heinemann, Oxford.

20. Honeycutt, J.D. and Andersen, H.C. (1987) Molecular Dynamics Study of Melting and Freezing of Small Lennard- Jones Clusters. The Journal of Physical Chemistry, 19, 4950-4963.

21. Tsuzuki, H., Branicio, P.S. and Rino, J.P. (2007) Structural Characterization of Deformed Crystals by Analysis of Common Atomic Neighborhood. Computer Physics Communications, 177, 518-523.

Chapter 4

Molecular Dynamic Study of Pull-In Instability of Nano-Switches

Sogand Hoshyarmanesh, Mohsen Bahrami*

Department of Mechanical Engineering, Amirkabir University of Technology, Tehran, Iran

ABSTRACT

Capacitive nano-switches have been of great interest as replacements for conventional semiconductor switches. Accurate determination of the pull-in voltage is critical in the design process. In the present investigation, pull-in instability of nano-switches made of two parallel plates subjected to electrostatic force is studied. For this purpose, two parallel rectangular nanoplates with opposite charges are modeled based on molecular dynamics (MD) technique. Different initial gaps between nanoplates and its effect on pull-in phenomena are studied in addition to taking different values of geometrical and physical parameters into account to evaluate pull-in voltages. Here molecular dynamic simulations as an atomic interaction approach are employed for modeling of nano-switches in order to study pull-in instability considering atomic interaction and surface tension. Boundary conditions and also the van der Waals force are considered as important parameters to investigate their effects on pull-in voltage values.

Keywords: Nanomechanics, Pull-In Instability, Electric Field, Molecular Dynamics Simulation, van der Waals Force

1. INTRODUCTION

The potential difference across the conductors develops an electric field across the dielectric, causing positive plate to come in contact with the ground plane. If the applied voltage exceeds this pull-in limit, the upper plate will snap down. Accurate determination of the pull-in voltage is critical in the design process of capacitive switches and in virtue of the electromechanical coupling effect and the nonlinearity of electrostatic force [1] [2]. To operate devices such as capacitors, they need to be biased with a DC voltage to form a surface charge [3] [4]. The aspects such as drive mode, temperature dependence and dielectric charging have been analyzed and their effects on the pulling-voltage are evaluated [5]. There are numerous investigations based on analytical modeling of the pull-in voltage for beams or diaphragms with the modified capacitance accounting to the fringing field and the equivalent spring constant to meet the emerging need for microelectromechanical systems (MEMS) process during both process development and manufacturing [6] -[11] . Distributed modeling based on well-known beam and plate theories [12] is also developed. Moreover, analytical approximate solutions for pull-in voltages are obtained via energy methods [13] , homotopy perturbation method [14] , modified couple stress theory [15] , and with elastic boundaries [16] . Parallel plate capacitors are modeled in the equations of motion based on classical elasticity theory and the orthodox theory of Coulomb blockade [17] Researchers have shown that there is a close match between analytical solution and numerical simulations of the nano-devises. For instance in [18] the continuum model's accuracy is compared with atomistic simulation to compute the pull-in voltage of Nano-cantilever switches.

Atomistic simulation is an applicable approach to studying nano system's properties [19]-[21]. Hoshyarmanesh et al. applied molecular dynamics to investigate surface molecular interaction in micro-cantilever biosensors [22]. Molecular dynamics simulation of pull-in phenomena is also derived in carbon nanotubes to study Stone-Wales defects, different geometries and boundary conditions on the pull-in charges [23] [24].

Different methods such as molecular mechanics and atomistic modeling are employed in order to investigate the mechanical behaviors of nanostructures. Ansari and Sahmani [25] studied biaxial buckling behavior of single-layered graphene sheets on the basis of two approaches including nonlocal continuum elasticity and MD simulation. Bahrami et al. used molecular dynamic to simulate the Interaction of Carbon Nanotubes and External Flow [26] .

Additionally, various investigations have been conducted to indicate the size effect on the pull-in phenomenon of microand nano-switches. Mohammadi et al. [27] investigated the size effect on the pull-in instability of hydrostatically and electrostatically actuated circular microplates. The research revealed that variation of dimensionless internal length scale parameter and hydrostatic pressure leads to different values of pull-in voltage. Mousavi et al. [28] studied the small scale effect on the pull-in instability of nano-switches subjected to electrostatic and intermolecular forces. The intermolecular and electrostatic attraction forces using Eringen's nonlocal elasticity theory are solved numerically by differential quadrature method. Wang and Wang [29] studied the pull-in instability of a nano-switch under electrostatic and intermolecular Casimir forces with consideration of the surface energy. The analysis is based on the geometrically non-linear Euler-Bernoulli beam theory with consideration of the surface energy. The research has proved the effect of intermolecular Casimir force on the pullin voltage weakens as the initial gap has increases.

Many problems in the natural sciences and engineering are common with sources of uncertainty. There are various numerical and analytical methods to solve the static uncertain systems like fuzzy finite element analysis [30] and other trial and errors approaches, however these are not approximate solutions and lack of accuracy is observable. Computer simulations are the most common approach to studying problems in uncertainty quantification. Simulation-based methods like Monte Carlo simulations are one of the basic probabilistic approaches for uncertainty propagation. Plates uniformly loaded with electrical potential and fixed at both four edges and two edges are studied here and also the nano-switche with fixedplate at four sides as an uncertain static system is solved through molecular dynamic technique.

In the current investigation, the pull-in instability of electrostatically actuated nano-switches is studied with considering the small scale effect conducting molecular dynamic simulation. To study pull-in phenomena nanoswitches with different initial gaps and different boundary conditions are simulated and pull-in voltages are evaluated. In addition to investigate the effects of van der Waals force in the small-scale, the simulation is solved by both considering and neglecting van der Waals force, and the results are compared.

2. MOLECULAR DYNAMICS SIMULATION

Two approaches have been used to obtain the solution of the maximum deflection for fixed thin rectangular plates under uniform load; these are the double cosine series and the superposition method as a generalization of Hencky's solution [31]. The problem of the uniformly loaded rectangular plate with fixed at all sides has been solved by Hencky (1913) and independently by Boobnoff (1902). Boobnoff made exact calculations for several aspect ratios for the plate while Hencky made refined calculations only for the case of a square plat [32]. Hutchinson has used the solution from which was presented in [33]. However molecular dynamic investigate the atomic interaction of plates in nano-devices considering non-local view.

Due to the difficulties in manufacturing nanostructure, experiments on nano-switches have been costly. Computational design tools, when sufficiently precise and accurate, can be far more economical than trial-

and-error design cycles based purely on experimentation. In this section, a molecular dynamic approach is proposed for analysis and design of capacitors. To this end, MD simulation of a nano-switch and it's response to electrical field and estimation of mechanical deflection of capacitor is performed. Schematic of nanoplate-based switch is shown in Figure 1. The capacitor plates are constrained with two different boundary conditions as are displayed in Figure 2, clamped square diaphragm and fixed-fixed sides, while the lower plate is fixed in bottom side.

The necessary step that must be considered in MD simulation is equilibration of the capacitor, since it is the starting point of minimization process [34]. NAMD execution consists of three stages of equilibration and one stage under applied electrical field [35]. The stages are prevalent in most of simulations. First stage is energy minimization; second stage temperature is increased from zero to 295 Kelvin degree with constant volume. In the next stage pressure is constant and the Langevin thermostat is applied. The last stage the volume is adjusted constant and the desire voltage is applied to system. The constant pressure simulation let volume to alter. An equilibrate system can be find out by following conditions, volume should fluctuate smoothly around a constant value; also energy is approached to a certain value. If one draws the volume curve versus time, a noticeable reducing tangent would be observed in the volume.

In order to simulate the mechanical deflection of a capacitor, the electrical field with different values is applied to the system. In other words, potential difference applied to a system causes a certain value of voltage along z-axis. Electrostatic forces deflect the moveable nanoplate depending on the gap between the movable plate and the fixed one. van der Waals force also acts an important role in the nanoplate's attraction. LennardJones potential has two attractive and repulsive terms. Forcefield potential energy function [36] consists of intermolecular or bonding interactions, Lennar-Jones potential for the vdW interaction and the Coulomb potential for electrostatic actuation

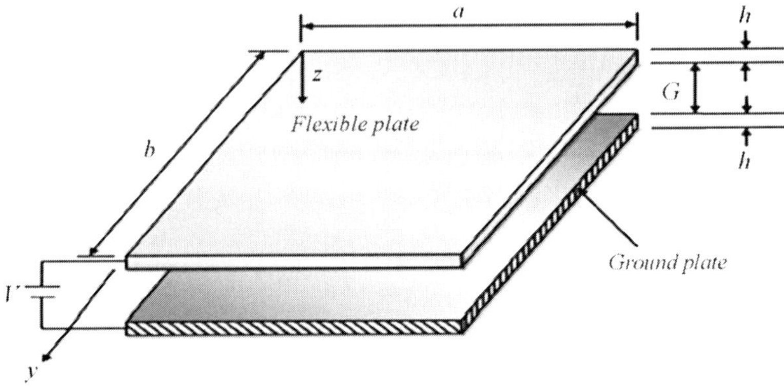

Figure 1. Schematic of nanoplate-based switch: kinematic parameters and coordinate system.

Figure 2. Boundary conditions (left) clamped-clamped (right) four sides fixed.

are utilized to model the nano-switches as it is brought in Equation (1) and expanded in Equation (2).

$$E = E_{bonded} + E_{vdW} + E_{coul} \tag{1}$$

$$E = \sum_{bonds} k_r (r - r_0)^2 + \sum_{angles} k_r (\theta - \theta_0)^2 + \sum_{dihedrals} k_\phi \left[1 + \cos(n\phi + \phi_0)\right] + \sum_{atoms\ i} \sum_{j \neq i} 4\varepsilon_{i,j} \left[\left(\frac{\sigma_{i,j}}{r_{i,j}} \right)^{12} - \left(\frac{\sigma_{i,j}}{r_{i,j}} \right)^{6} \right] + \sum_i \sum_{j \neq i} \frac{q_i q_j}{\varepsilon_0 r_{i,j}} \tag{2}$$

where $r_{i,j}$ is the distance between atoms i and j, the parameter ε denotes the strength of the interaction and σ defines a length scale. The attractive term proportional to $r_{i,j}^{-6}$ dominates at large distances and it models the van der Waals dispersion forces caused by the dipole-dipole interactions due to fluctuating dipoles. The electrostatic interaction of two atoms with charges q_i and q_j separated by $r_{i,j}$ is given by E_{coul}, the dielectric contant ε_0 in vacuum is equal to 1. When two plates contact with each other, the vdW repulsive force will make the collapse harder; the total van der Waals energy can be computed by a pairwise summation over all the atoms.

Plats are made of silicon which the upper plate is n-type silicon and the lower one is p-type silicon. This basically means n-type silicon is positively charged, and p-type silicon becomes negatively charged. Subsequently applying an uniform electric field across the z-axis enforce an attractive force between plats and voltage has a direct relation to electric field, which is electric field multiple the size of system along the z-axis, consequently the pull-in voltage will be determined.

When an electrical field is applied, the lines of pressure control will be absent. Inserting an electrical field with controlled pressure disturbs the system and led to incorrect answers. In addition the Langevin temperature controlled should be turning on. Applying Langevin forces to ions, their motions which supposed to be measured through applied electrical field, led to an infinitesimal deviation in the current. If a high electrical field was applied to system, the high ionic current would increase the temperature up to 450 Kelvin degree. The simulation leans to be invalid in such degree of heat, since this restriction limits us to work by smaller electrical field in real simulation. In the MD simulation, the atom's interaction and the surface tension are considered that impose an ineligible force on nano-scale devices. Surface tension plays an important role in nano-devices function which is neglected in classic mechanic. It makes the simulation more accurate than analytical calculation such as continuum mechanics that is originated from classic mechanics.

Total number of atoms simulated to form plates is 66,304, and plates are made of silicon. Force field parameters of silicon for the covalent bonds between atoms including bond energy, bond radius, angle, dihedral, van der Waals radius and some other decisive factors are brought in theAppendix. Overall simulation time is determined based on reaching to a stable point in energy and volume, as it is seen in following graphs the simulation steps has exceeded 120,000 steps considering time step fixed at 2 fs/step. Fixed atom constraint facility made it possible to take the effects of boundary condition into account.

Simulation is adjusted as the coordinate of atoms be registered once in every 100 steps to find the pull-in voltage. As it is illustrated in the Figures 3-5 the applied electric field should be large enough to contact plates.

The capacitor with gap value of 2.719 nm was actuated by the electric field of 7 $\frac{\text{kcal}}{\text{mol}\cdot \text{A}^{\circ}\,\text{e}}$ and 5 $\frac{\text{kcal}}{\text{mol}\cdot \text{A}^{\circ}\,\text{e}}$ considering with and without vdW respectively. The electric field value is gradually augmenting for the gap values of 1.543, 4.3504 nm, respectively, in the case of c lamped-clamped boundary condition. Thus, absorption has happened over the time and plates have connected completely.Figure 5 demonstrates plate's absorption in the case of four sides fixed boundary condition regarding gap value of 4.3504.

The Newtonian equations of motion are solved for the atoms which are consisting of following terms[37] :

$$r_i^{\cdot} = \frac{P_i}{m_i}, \qquad P_I^{\cdot} = F_i + q_i E_i \tag{3}$$

where E_i is the electric field acting on atom i, m_i the mass, q_i the partial charge, and pi the laboratory momentum of atom i. F_i represents the total intermolecular force acting on atom i due to all other atoms.

Silicon mass and partial charge are respectively equal to 28 u and +4 for n-type and −4 for p-type.

Simulation has been executed with and without van der Waals force to discover the exact effect of surface forces in the function nano-scale switches. Molecular dynamic results indicate that the der Waals force can play a significant role in determining the pull-in voltage of cantilever switches especially for smaller gaps. When the gap between the cantilevers plats is very small, vdW force acts as a repelling force to avoid the distance to exceed the atomic van der Waals radius. Under this circumstance system requires a bigger snap-down voltage to overcome the repulsive force.

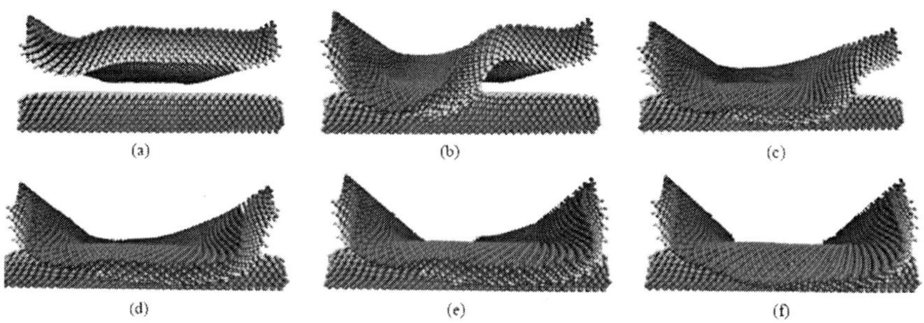

Figure 3. MD simulation performed for pull-in phenomenon in a nano-switch, fixed at two sides, gap = 4.3504.

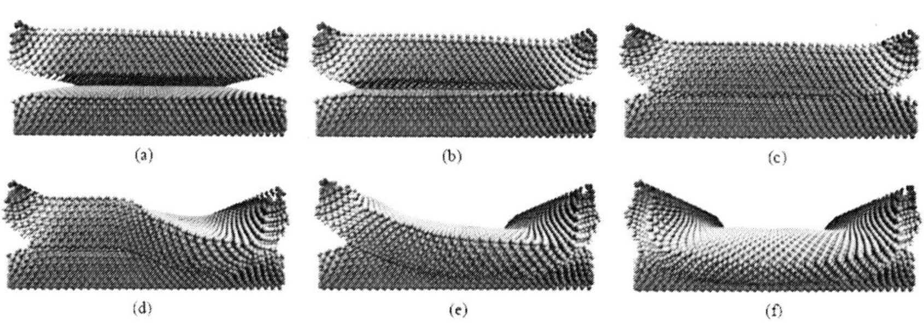

Figure 4. MD simulation performed for pull-in phenomenon in a nano-switch, fixed at two sides, gap = 1.543.

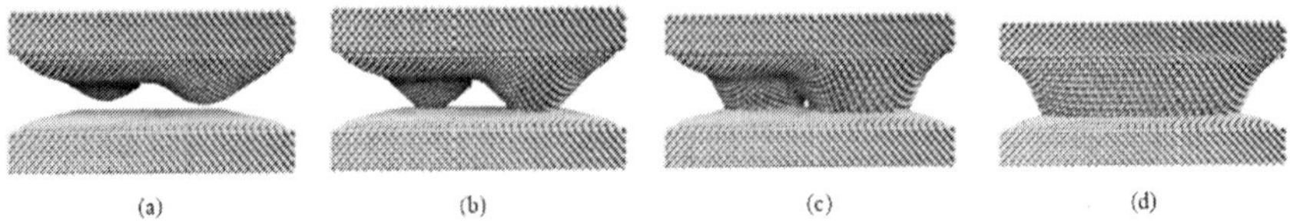

Figure 5. MD simulation performed for pull-in phenomenon in a nano-switch, fixed at four sides, gap = 4.3504.

3. RESULTS AND DISCUSSION

The successful application of molecular dynamic simulations is using a nonlinear least square fitting procedure to minimize the equilibrate energy and obtain stabilize system results. The geometrical parameters of nanoswitches are considered in MD simulations, i.e. $h = 2.17$ nm, $a = b = 15.07$ nm as are seen below. Capacitors with two different gaps are modeled in Figure 3, Figure 4 and plates are fixed at two sides; however, in Figure 5 upper plate is fixed at four sides. The figures show the transient absorption of plates under the electrical field so the whole parts are illustrating the snap-down phenomena. Although for pulling phenomenon just a small connection between the electrodes are enough, as shown in the Figures 3-5, the connection is not so small and a remarkable area of the upper electrode touches the lower electrode, this is due to gaps size designed in the vicinity of plats thickness value in addition to a potent atomic interaction. In the other word relatively small gaps, which are in the order of plats thickness, make a large area of plats surface to contact.

The values of pull-in voltages obtained by MD simulations are tabulated in Table 1 corresponding to nanoswitches with different initial gaps between the nanoplates. It is clearly seen that by increasing the value of initial gap of nano-switch, the pull-in instability occurs at higher applied voltages.

As a necessary condition for equilibration, the volume should be smoothly fluctuated around an average value and stabilized energy lean to a certain value. Figure 6 and Figure 7 demonstrate the system's steady state energy versus time-step regarding four different initial capacitor's gaps and fixed edges. As it is demonstrated the system's total energy has remained constant during equalization. If the gap between the upper electrode and the ground plane is smaller than vdW radius, van der Waals force acts inversely, so applies an repelling force that inhibit plates from having contact; to obtain precious results vdW must be taken into account. Numerical value of vdW force during equalization is drawn distinguishingly in Figure 8. One can see that the graph reaches its peak in the plats distance of 0.22 nm before dropping and converting to a repellent force.

To have a better judgment, graphs of applied voltage in fourvarious gaps are drawn in one picture.Figure 9 illustrates the applied voltages versus deflection regarding different gap values; in addition pull-in voltages being obtained by the MD simulations versus their relevant gap value are drawn as a separated curve. It is indicated that the plats initial distance as a part of geometry plays an important role in the pull-in phenomenon of nano-switches.

The pull-in voltage versus gap values in case of investigation of boundary condition (fixed plates at two sides and four sides) and vdW effects are respectively illustrated in Figure 10 and Figure 11which present a nonlinear relation of them. In the Figure 11 one can clearly perceive the obtained pull-in voltage is directly influenced by the van der Waals force; for a more intensive contact the repulsive force must be defeated, so the plats are so close, it requires a bigger electric field to frustrate vdW repulsive force. To emphasize this concept nano-switch sensitivity is calculated and shown in Figure 12 in two cases of considering and ignoring vdW based on following formula, output(deflection) divided input(actuation voltage), also the boundary condition is considered into sensitivity. To increase the sensitivity and decrease the pull-in voltage it is viable to increase the degree of freedom of plates in addition to decreasing the gap value.

Table 1. The values of pull-in voltages predicted by MD simulations corresponding to various initial gaps and boundary conditions.

Initial gap (nm)	Pull-in voltage (v) —clamped-clamped	Pull-in voltage (v) —clamped-clamped—vdW	Pull-in voltage (v)—four sides fixed
0.514	0.2118	0.465	0.4707
1.543	1.1763	2.3536	5.8841
2.72	15.3054	21.4275	79.5879
4.3504	60.2962	64.0647	177.1202

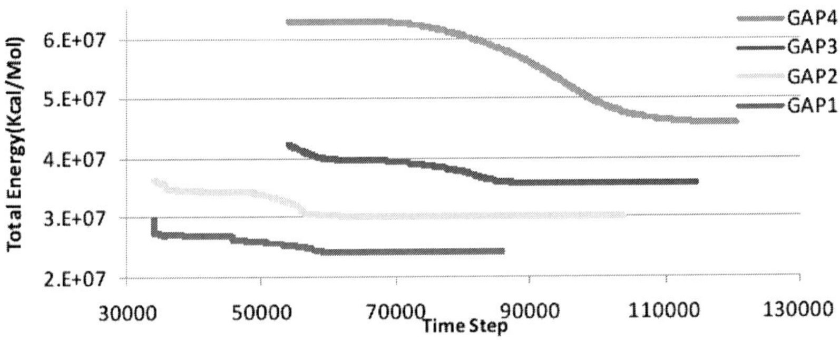

Figure 6. Variation of system's total energy versus step time regarding various initial gaps, fixed at two sides, G1 = 0.514 nm, G2 = 1.08 nm, G3 = 2.72 nm, G4 = 4.3504 nm.

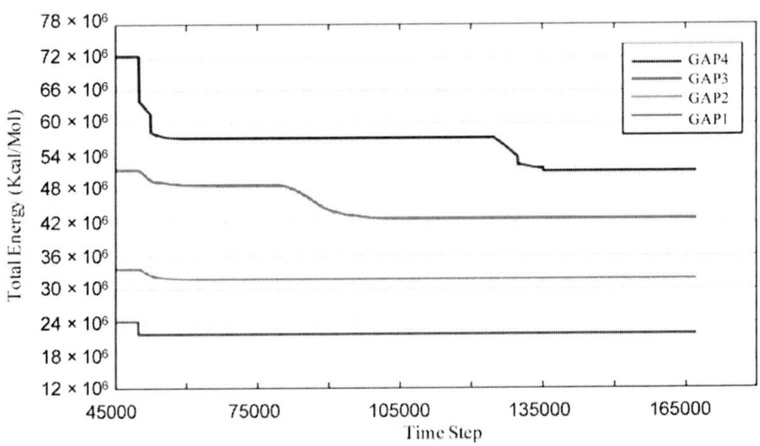

Figure 7. Variation of system's total energy versus step number regarding various initial gaps, fixed at four sides.

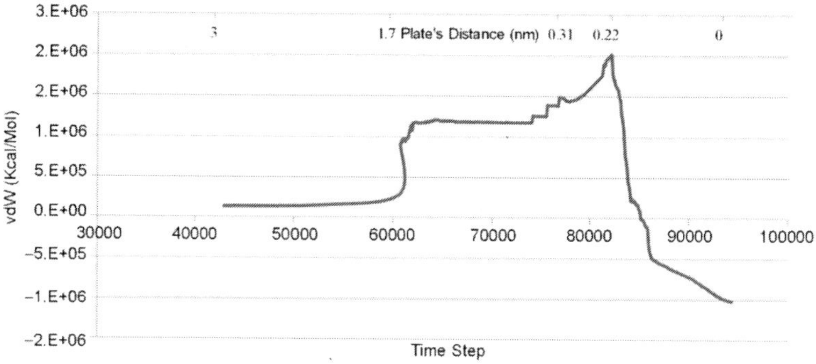

Figure 8. Variation of system's van der Waals energy versus step number for silicon plats, fixed at two sides.

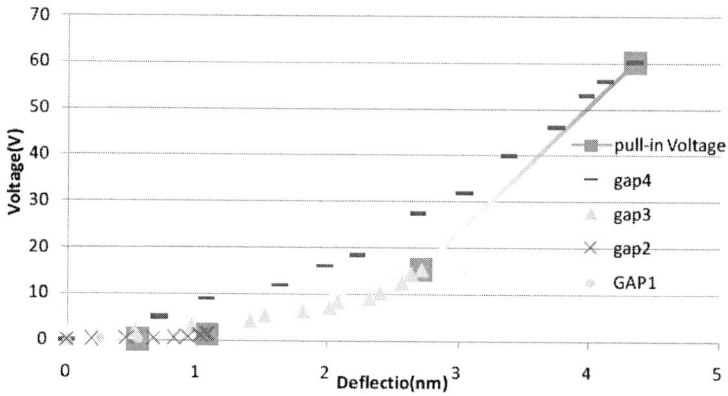

Figure 9. Variation of pull-in voltage versus deflection under clamped-clamped boundary condition regarding various gap values as: G1 = 0.514 nm, G2 = 1.08 nm, G3 = 2.27 nm, G4 = 4.3504 nm.

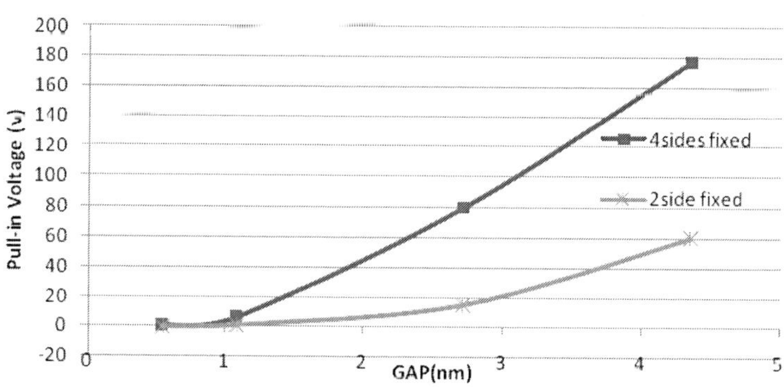

Figure 10. Variation of pull-in voltage versus gap value regarding boundary conditions.

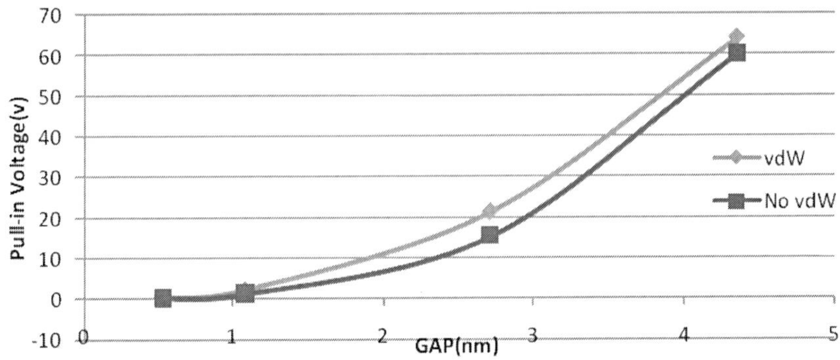

Figure 11. The pull-in voltage versus gap values considering vdW effects.

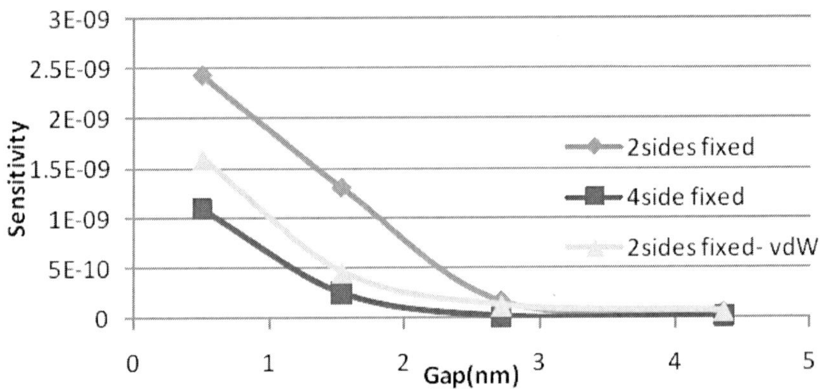

Figure 12. Variation of sensitivity versus gap value.

4. CONCLUSION

In the present work, the pull-in phenomenon in nano-switches considering gap value and boundary condition is investigated. MD simulations as a unique view in simulation of pull-in phenomena are performed in capacitor switches with different boundary conditions and gap values. Pull-in voltages of nano-switches are predictedus ing MD. It was revealed that plats' initial gap has a considerable influence on the pull-in phenomenon in nanoswitches. van der Waals force plays a significant role in plate's absorption especially in the case of smaller gaps, based on the results that the effects of van der Waals force have intensified in nano-scaled switches and played as an auxiliary force at first; however, for closer contact its role has changed to a repulsive force. Clamped edges compel a resisting force against electrical field which acquires system an extra load for plates contact, hence the pull-in voltage is much bigger in the case of four sides fixed plates in comparison to two sides fixed. Mechanism of pull-in phenomena is analyzed and studied without carrying out time-consuming and high cost experiments; through molecular dynamic it is workable to study each important parameter thoroughly. That is the reason that makes the MD a powerful technique to scrutinizenano-devices and their internal interaction like force field potential, electrostatic forces and van der Waals forces.

REFERENCES

1. Hu, Y.-C., Chang, C. and Huang, S. (2004) Some Design Considerations on the Electrostatically Actuated Microstructures. Sensors and Actuators A: Physical, 112, 155-161.
2. Hu, Y.-C., Wang, Y.H. and Huang, S.C. (2004) On the Dynamic and Stability Analysis of Electrostatically Actuated Microstructures. Proceedings of the 1st International Symposium on Micro & Nano Technology (ISMNT-1), III, 2-01.
3. Pappalardo, M. and Caronti, A. (2002) A New Alternative to Piezoelectric Transducer for NDE and Medical Applications: The Capacitive Ultrasonic Micromachined Transducer (cMUT). University Roma, Rome.
4. Pappalardo, M. (2002) A New Approach to Ultrasound Generation: The Capacitive Micromachined Transducers. University Roma, Rome.

5. Rocha, L., Cretu, E. and Wolffenbuttel, R. (2004) Full Characterisation of Pull-In in Single-Sided Clamped Beams. Sensors and Actuators A: Physical, 110, 301-309.

6. Osterberg, P.M. and Senturia, S.D. (1997) M-TEST: A Test Chip for MEMS Material Property Measurement Using Electrostatically Actuated Test Structures. Journal of Microelectromechanical Systems, 6, 107-118.

7. Pamidighantam, S., et al. (2002) Pull-In Voltage Analysis of Electrostatically Actuated Beam Structures with FixedFixed and Fixed-Free end Conditions. Journal of Micromechanics and Microengineering, 12, 458-464.

8. Gupta, R.K. (1998) Electrostatic Pull-In Test Structure Design for In-Situ Mechanical Property Measurements of Microelectromechanical Systems (MEMS), Citeseer.

9. Wei, L.C., Mohammad, A.B. and Kassim, N.M. (2002) Analytical Modeling for Determination of Pull-In Voltage for an Electrostatic Actuated MEMS Cantilever Beam. IEEE International Conference in Semiconductor Electronics, 19-21 December 2002, 233-238.

10. Ganji, B.A. (2012) Accurate Determination of Pull-In Voltage for MEMS Capacitive Devices with Clamped Square Diaphragm. International Journal of Engineering, 25, 161-166.

11. Rocha, L., Cretu, E. and Wolffenbuttel, R. (2003) Displacement Model for Dynamic Pull-In Analysis and Application in Large-Stroke Electrostatic Actuators. Proceedings of Eurosensors 17, 17th European Conference on Solid-State Transducers, Guimar, 20-21 September 2003, 448-451.

12. Osterberg, P., Yie, H., Cai, X., White, J. and Senturia, S. (1994) Self-Consistent Simulation and Modelling of Electrostatically Deformed Diaphragms. IEEE Workshop on Micro Electro Mechanical Systems, MEMS'94, Oiso, 25-28 January 1994, 28-32.

13. Hu, Y.C. and Lee, G.D. (2007) A Closed Form Solution for the Pull-In Voltage of the Micro Bridge Tamkang. Journal of Science and Engineering, 10, 147-150.

14. Fakhrabadi, M.M.S., Rastgoo, A. and Ahmadian, M.T. (2013) Analysis of Pull-In Instability of Electrostatically Actuated Carbon Nanotubes Using the Homotopy Perturbation Method. Journal of Mechanics of Materials and Structures, 8, 385-401.

15. Fakhrabadi, M.M.S., Rastgoo, A., Ahmadian, M.T. and Mashhadi, M.M. (2014) Dynamic Analysis of Carbon Nanotubes under Electrostatic Actuation Using Modified Couple Stress Theory. Acta Mechanica, 225, 1523-1535.

16. Hu, Y.C., Chang, P.Z. and Chuang, W.C. (2007) An Approximate Analytical Solution to the Pull-In Voltage of a Micro Bridge with an Elastic Boundary. Journal of Micromechanics and Microengineering, 17, 1870-1877.

17. Kinaret, J.M., Nord, T. and Viefers, S. (2003) A Carbon-Nanotube-Based Nanorelay. Applied Physics Letters, 82, 1287-1289.

18. Dequesnes, M., Rotkin, S. and Aluru, N. (2002) Calculation of Pull-In Voltages for Carbon-Nanotube-Based Nanoelectromechanical Switches. Nanotechnology, 13, 120.

19. Istadeh, K.H., Kalantarinejad, R., Aghaei, M.J. and Yazdi, M.R.S. (2013) Computational Investigation on H2S Adsorption on the CNT Channel of Conductometric Gas Sensor. Journal of Computational and Theoretical Nanoscience, 10, 2708-2713.

20. Sarkisov, L. and Harrison, A. (2011) Computational Structure Characterisation Tools in Application to Ordered and Disordered Porous Materials. Molecular Simulation, 37, 1248-1257.

21. Smith, W., Yong, C. and Rodger, P. (2002) DL_POLY: Application to Molecular Simulation. Molecular Simulation, 28, 385-471.

22. Hoshyarmanesh, S., Bahrami, M. and Kalantarinejad, R. (2014) A Multiscale Approach in the Computational Modeling of Bio-Physical Environment of Micro-Mechanical Biosensor towards the Prostate Specific Antigen Diagnosis. Journal of Computational and Theoretical Nanoscience, 11, 1374-1384.

23. Fakhrabadi, M.M.S., Khorasani, P.K., Rastgoo, A. and Ahmadian, M.T. (2013) Molecular Dynamics Simulation of Pull-In Phenomena in Carbon Nanotubes with Stone-Wales Defects. Solid State Communications, 157, 38-44.

24. Fakhrabadi, M.M.S., Rastgoo, A. and Ahmadian, M.T. (2013) On the Pull-In Instability of Double-Walled Carbon Nanotube-Based Nano Electromechanical Systems with Cross-Linked Walls. Fullerenes, Nanotubes and Carbon Nanostructures.

25. Ansari, R. and Sahmani, S. (2013) Prediction of Biaxial Buckling Behavior of Single-Layered Graphene Sheets Based on Nonlocal Plate Models and Molecular Dynamics Simulations. Applied Mathematical Modelling, 37, 7338-7351.

26. Bahrami, M., Kalantarinejad, R., Aghaei, M.J. and Azadi, N. (2011) Simulation of the Interaction of Carbon Nanotubes and External Flow. Journal of Computational and Theoretical Nanoscience, 8, 563-567.

27. Mohammadi, V., Ansari, R., Shojaei, M.F., Gholami, R. and Sahmani, S. (2013) Size-Dependent Dynamic Pull-In Instability of Hydrostatically and Electrostatically Actuated Circular Microplates. Nonlinear Dynamics, 73, 1515-1526.

28. Mousavi, T., Bornassi, S. and Haddadpour, H. (2013) The Effect of Small Scale on the Pull-In Instability of NanoSwitches Using DQM. International Journal of Solids and Structures, 50, 1193-1202.

29. Wang, K.F. and Wang, B.L. (2014) Influence of Surface Energy on the Non-Linear Pull-In Instability of NanoSwitches. International Journal of Non-Linear Mechanics, 59, 69-75.

30. Behera, D. and Chakraverty, S. (2013) Fuzzy Finite Element Based Solution of Uncertain Static Problems of Structural Mechanics. International Journal of Computer Applications, 59, 69-75.

31. Wojtaszak, I.A. (1937) The Calculation of Maximum Deflection, Moments, and Shear for Uniformly Loaded Rectangular Plate with Clamped Edge. Journal of Applied Mechanics, Transactions ASME, 4, A173-A176.

32. Szilard, R. (1973) Theory and Analysis of Plates: Classical and Numerical Methods. Prentice-Hall, Upper Saddle River.

33. Timoshenko, S., Woinowsky-Krieger, S. and Woinowsky, S. (1959) Theory of Plates and Shells. Vol. 2, McGraw-Hill, New York.

34. Hansson, T., Oostenbrink, C. and van Gunsteren, W. (2002) Molecular Dynamics Simulations. Current Opinion in Structural Biology, 12, 190-196.

35. Phillips, J.C., Braun, R., Wang, W., Gumbart, J., Tajkhorshid, E., Villa, E., et al. (2005) Scalable Molecular Dynamics with NAMD. Journal of Computational Chemistry, 26, 1781-1802.

36. Verlet, L. (1967) Computer "Experiments" on Classical Fluids. I. Thermodynamical Properties of Lennard-Jones Molecules. Physical Review, 159, 98.

37. De Luca, S., Todd, B.D., Hansen, J.S. and Daivis, P.J. (2013) Electropumping of Water with Rotating Electric Fields. The Journal of Chemical Physics, 138, Article No. 154712.

Chapter 5

Variability and Reliability of Single-Walled Carbon Nanotube Field Effect Transistors

Ahmad Ehteshamul Islam [1,2]

[1]*Soft Matter Materials Branch, Materials and Manufacturing Directorate, Air Force Research Laboratory, Wright-Patterson Air Force Base, Dayton, OH 45433, USA*

[2]*Biological and Nanoscale Technologies, UES Inc., Dayton, OH 45432, USA*

ABSTRACT

Excellent electrical performance and extreme sensitivity to chemical species in semiconducting Single-Walled Carbon NanoTubes (s-SWCNTs) motivated the study of using them to replace silicon as a next generation field effect transistor (FET) for electronic, optoelectronic, and biological applications. In addition, use of SWCNTs in the recently studied flexible electronics appears more promising because of SWCNTs' inherent flexibility and superior electrical performance over silicon-based materials. All these applications require SWCNT-FETs to have a wafer-scale uniform and reliable performance over time to a level that is at least comparable with the currently used silicon-based nanoscale FETs. Due to similarity in device configuration and its operation, SWCNT-FET inherits most of the variability and reliability concerns of silicon-based FETs, namely the ones originating from line edge roughness, metal work-function variation, oxide defects, *etc.* Additional challenges arise from the lack of chirality control in as-grown and post-processed SWCNTs and also from the presence of unstable hydroxyl (–OH) groups near the interface of SWCNT and dielectric. In this review article, we discuss these variability and reliability origins in SWCNT-FETs. Proposed solutions for mitigating each of these sources are presented and a future perspective is provided in general, which are required for commercial use of SWCNT-FETs in future nanoelectronic applications.

Keywords: single-walled carbon nanotube; field effect transistor; variability; reliability; hydroxyl group passivation; oxide defect; hysteresis; noise; radiation dose; degradation

1. INTRODUCTION

Single-Walled Carbon NanoTubes (SWCNTs) were first discovered in 1991 [1] and their electronic properties depend on the directions at which two-dimensional graphene stripes are (hypothetically) rolled to make SWCNTs [2,3,4]. When a SWCNT is defined by rolling a graphene layer and by joining any of the solid symbols of Figure 1a with the lattice origin (a_1,a_2) = (0,0), resultant SWCNT will have metallic properties with no bandgap in the energy *vs.* density of the state's relationship (Figure 1b). On the other hand, SWCNTs defined by joining the open symbols of Figure 1a with the lattice origin will have semiconducting properties with bandgaps (Figure 1c,d) that can be expressed as [2,3]:

$$E_G = \frac{2a_{C-C}|t|}{d}. \quad (1)$$

In Equation (1), d is the diameter of SWCNT and equals $\sqrt{3}a_{C-C}/\pi \times \sqrt{(n^2 + nm + m^2)}$, (n,m) is the chiral index designating the distance between the open symbols and the lattice origin via chiral vector $\bar{c} = n\bar{a_1} + m\bar{a_2}$ (Figure 1a), a_{C-C} = 0.142 nm is the C–C bond length, t ~ −3 eV is the C–C bonding energy. SWCNTs with $n = m$ are called armchair nanotubes and always show metallic behavior, whereas SWCNTs with $n \neq m$ either show metallic behavior for mod $(n - m)/3 = 0$ or semiconducting behavior for mod $(n - m)/3$ = 1, 2. SWCNTs with $m = 0$ are called zigzag nanotubes and the rest with $n \neq m \neq 0$ are called chiral nanotubes.

Semiconducting SWCNTs (s-SWCNTs) are of particular interest for future electronic devices, where s-SWCNTs act as a channel in conventional field effect transistor (FET) configuration. Carrier mobility in such s-SWCNT-FETs can be as high as ~10^5 cm^2/V-s [5,6], which outperforms similar values in current-day FETs. In addition, the ultra-thin s-SWCNT layer that forms the channel of FET offers better electrostatic gate control, hence less short channel effect compared to classical silicon-based FETs [7]. Such ultra-thin layers also offer extreme mechanical flexibility to the resultant electronic device, thereby promoting the best platform for future flexible electronic devices [8,9], which are currently being manufactured using FETs made on thin silicon substrates [10,11]. Use of s-SWCNTs as channels, m-SWCNTs as contacts [12] and memory elements [13], SWCNT network [14,15] or single s-SWCNT [16,17] as memory, and CNT fabrics as coaxial cable [18] may open up the possibility of having “all-CNT” electronics for high performance applications. In addition to electronics, SWCNT-FETs are also considered as chemical-/bio-sensors because of their higher surface-to-volume ratio that reduces response time [19] and extreme sensitivity [20,21], and their selectivity [20] to chemical and biological species.

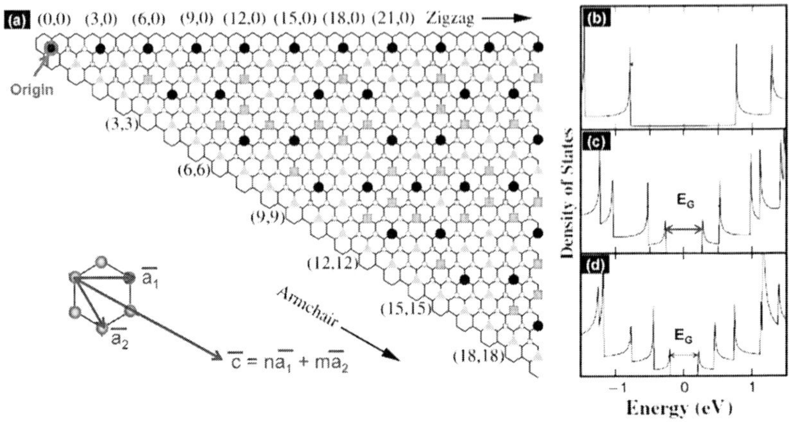

Figure 1. (**a**) Two-dimensional graphene lattice for defining structures of (*n*,*m*) Single-Walled Carbon NanoTubes (SWCNTs) (reprinted with permission from Ref. [3]; © 2005 by the American Physical Society.). Definition of lattice vectors a_1, a_2 with respect to carbon atoms in the graphene lattice is shown in the bottom part of the figure. SWCNTs can be formed by rolling this sheet to join the origin (a_1,a_2) = (0,0) with other points. Joining the open symbols with the origin makes semiconducting SWCNTs, while joining the solid symbols with the origin makes metallic SWCNTs. (**b**) Density of states (DOS) *vs.* energy plot obtained using a tight binding model [4] for an armchair (10, 10) SWCNT having metallic property. (**c**,**d**) DOS +energy plot for a chiral (11, 9) SWCNT and a zigzag (22, 0) SWCNT. Both these SWCNTs have semiconducting properties and a chirality-dependent band gap (E_G) defined using Equation (1). (**b**–**d**) are reprinted with permission from Ref. [4]; © 2003 IOP Publishing Ltd and Deutsche Physikalische Gesellschaft, Published under a CC BY-NC-SA license.

To enable all these applications stated in the last paragraph, SWCNT-FETs and memories not only require superior performance, but also require a controlled level of variability and reliability. Such controllability has to ensure that performance of associated devices does not drop below a certain limit, which can be compensated through circuit design via guard bands [22,23]. Figure 2 schematically explains the extent of guard band that is used in circuit design for handling variability and reliability. Devices have different variability sources that give rise to time-zero variation in device performance around its mean value. In addition, during the operation of the device, its performance degrades over time and such degradation comes with its own statistical variation. Based on this knowledge, circuits are designed with such devices considering that the tail of the performance distribution at the end of intended lifetime will not go beyond a certain limit—thus, the designed circuits are guard-banded against performance variations. As device performance improves via Moore's law scaling [24], device design is continuously modified to minimize guard-band such that the devices can be operated at higher performance without compromising the lifetime.

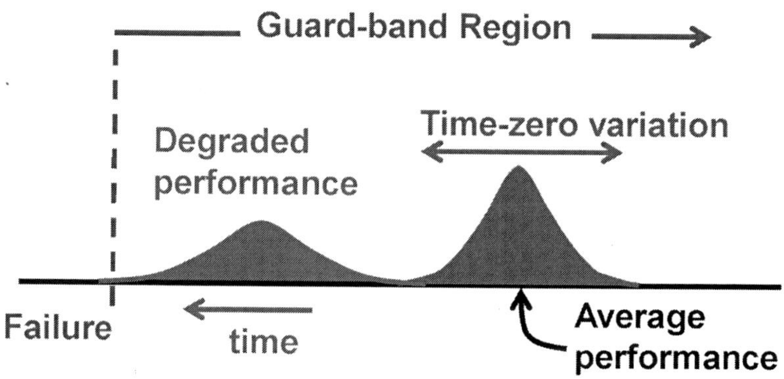

Figure 2. Schematic representation of the amount of guard-band that is used in electronics for handling process (or time-zero) variation and reliability (time dependent variation) in device performance. The performance of a device has a distribution over and above its mean value and it degrades over time. Once of the tail of performance distribution for the device reaches the guard-band margin, the associated device is designed to fail.

In this review article, we will discuss different aspects of variability and reliability in SWCNT-FETs that will eventually define guard-band in such devices. Section 2 discusses the configurations and performance of SWCNT-FETs that has potential use in future electronics. Section 3 highlights the sources of variability in SWCNT-FETs and the current solutions to control them. Section 4 performs a similar review for the reliability sources. Section 5 summarizes the paper and provides brief guidelines for improving variability and reliability phenomena in SWCNT-FETs based on current understanding of CNT growth conditions and device fabrication challenges.

2. CONFIGURATIONS AND PERFORMANCE OF SWCNT-FET

2.1. Device Geometry

For FET applications, SWCNTs are commonly grown in random networks or aligned arrays using metal catalyst placed on SiO_2 [9,12,25,26]. Alternately, SWCNTs are drop-casted from solution on SiO_2 [27,28,29] or hafnium dioxide (HfO_2) [30,31] and assembled in random networks [27,28,29] or parallel arrays [31,32] via coating of self-assembled monolayers on oxides. Bottom-gated FETs with thick SiO_2 as gate dielectric and heavily doped silicon as gate are widely used test configurations for SWCNT-FETs. Such bottom-gated structures though are suitable for bio-/chemical-sensing applications [20,21,33], top-gated geometry with thin, high-κ gate dielectric is preferred for high performance electronics [34,35]. (In some novel SWCNT-FETs, local bottom-gating [36], lateral gating with suspended SWCNTs [37], and surround-gate geometries [38] are also used for high performance electronics.) Measurement of SWCNT properties (e.g., conductivity, mobility, contact resistance of individual SWCNTs and their diameter dependence) by applying suitable bias at source, drain, and gate terminals involves fabrication of single SWCNT-FET. Schematic of such a FET in dual-gate (*i.e.*, having top and bottom gate with suitable gate oxides) configuration is shown in Figure 3a. Such single SWCNT-FET, however, cannot provide the amount of current required for different applications, e.g., ~mA per μm of channel width is required for high performance electronics [35]. Current in SWCNT-FETs can be increased either by having a network (Figure 3b) or an aligned array (Figure 3c) of SWCNTs' bridging sources and drain terminals. Magnitude of current in these configurations depends on the density of SWCNTs (per unit area for network SWCNTs ofFigure 3b or per unit width across the length of SWCNTs for aligned SWCNTs of Figure 3c), diameter of SWCNTs, chirality of SWCNTs, and electrostatic coupling on SWCNTs via gate. For network SWCNT-FETs, current additionally depends on the charge transfer efficiency across different SWCNTs in the network [39,40,41].

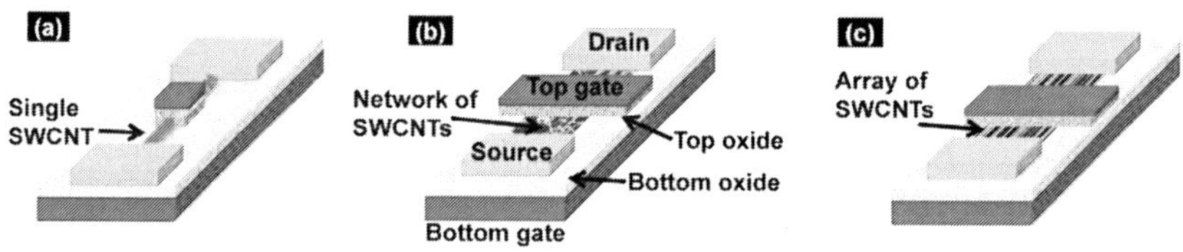

Figure 3. Schematic representation of SWCNT-based field effect transistor (FETs) in top- and bottom-gated configuration for (**a**) single SWCNT, (**b**) SWCNT networks, and (**c**) aligned array SWCNTs. Generally, SWCNTs are grown or coated on oxides and then the source, drain, top oxide and top gate are formed on top of them via lithography. Though the schematics here show partial coverage of SWCNTs in between source and drain, the actual top-gate SWCNT-FET application uses full coverage of SWCNTs.

2.2. Materials for SWCNT-FETs

As mentioned before, SWCNTs are grown on insulating substrates like SiO_2, quartz, sapphire, *etc.* using metal (e.g., Fe, Ni, Co and their alloys) nanoparticles as catalyst. Length, diameter distribution, and chirality of grown SWCNTs vary among the growth methods being used [42,43]. In general, for as-grown SWCNTs, length varies from ~1 µm to several 100 µm, diameter ranges between ~0.7 nm and ~5 nm [26,44,45,46], density is approximately 14–200 per µm^2 for network SWCNTs [12,31] and ~2–45 per µm across the length of the SWCNT for aligned SWCNTs [26,31,44,47,48], and chirality varies randomly. Long SWCNTs (>100 µm) are only obtained during chemical vapor deposition (CVD) on stable temperature- (ST-) cut quartz substrates [25,49]. Diameter and chirality of as-grown SWCNTs can be sorted by dissolving SWCNTs in surfactants and then using post-processing techniques like ultra-centrifugation [50], gel electrophoresis [51], chromatography [52]. Such solution-processed, sorted SWCNTs can be assembled into a high-density compact film via Langmuir-Blodgett method [35]. However, solution-processing breaks SWCNTs into small segments (~µm or less) and reduces the carrier mobility or conductivity of SWCNTs (down to ~1–20 cm^2/V-sec [35,46] with a maximum of ~200–300 cm^2/V-sec [28,53]). An alternative approach of maintaining the integrity of SWCNTs in terms of length (up to several hundreds of µm) and carrier mobility (>1000 cm^2/V-sec) is to coat horizontally aligned SWCNTs with a molecular film and use thermo-lithography (thermal analogous of photo-lithography) to etch-off m-SWCNTs from the substrate [54]. Remaining s-SWCNTs have a low density; however, there is potential for density improvement via multiple transfers [55,56].

In addition to the requirement of having substrates with high density s-SWCNTs, future electronic devices require appropriate doping in s-SWCNTs to control their conduction. Absence of molecular doping in SWCNTs, however, necessitates the use of metals with lower (higher) work-functions to conduct electrons (holes) through the SWCNTs, hence make n- (p-) type SWCNT-FETs. Palladium (Pd) with work-function of ~5.1 eV is generally used as source/drain contacts for making p-type SWCNT-FETs that allows hole flow from contacts into and out of the SWCNT via Schottky contacts for $d < 1$ nm SWCNTs and via ohmic contact for $d > 1$ nm SWCNTs [57,58,59]. Presence of negatively charged hydroxyl (-OH) groups at SWCNTs interface also makes the p-type conduction in SWCNT-FETs easier. N-type or electron conduction in SWCNTs, on the other hand, can be enabled using calcium (Ca) and scandium (Sc) source/drain contacts with work-functions ~3–4 eV [60,61]. In such case, conduction can be further enhanced via electrostatic doping with high-κ oxides [38,62].

2.3. FETs with Single SWCNT

Current-voltage (I-V) characteristics of FETs with single s-SWCNT have reasonably good switching behavior in the measured transfer characteristics (*i.e.*, drain current (I_{ds}) *vs.* gate voltage (V_g) characteristics, as shown inFigure 4a) and good saturation in output characteristics (*i.e.*, I_{ds} *vs.* drain to source voltage (V_{ds}) characteristics, as shown in Figure 4b) for any channel length (L_{ch}) down to ~10 nm [63]. These FETs have predominant p-type behavior because of the natural presence of –OH groups at the interface of SWCNTs. On-current (I_{on}; defined asI_{ds} at a certain high $|V_g - V_t|$, where V_t is the threshold voltage of a FET) and off-current (I_{off}; defined as the minimum of I_{ds}) in these FETs with single s-SWCNT show variation with d and source/drain contact material being used. Contacts with higher work-function (Φ_{WF}) can inject holes into the valence bands of s-SWCNTs more easily and have larger I_{on} and I_{off} (Figure 4c,d); [44,58,59]), especially for larger diameter s-SWCNTs that have smaller band-gap, E_g (see Equation (1)) and hence have negligible hole injection barrier at the metal/s-SWCNT interface. Therefore, contacts play a major role in the conduction of SWCNT-FETs and are considered to have dominant contribution in limiting I_{ds} in the near ballistic limit [28,63], when L_{ch} is much less than the channel mean free path that defines mobility. Sub-threshold slope (*SS*; defined as $\partial V_g/\partial(log_{10}I_{ds})$ in the region with $V_g << V_t$) for FETs with single s-SWCNT varies from ~80–95 mV/dec and increases with a decrease in L_{ch} due to ambipolar conduction [64]. (Use of appropriate chemical coating on SWCNT in a dual-gate device configuration can reduce *SS* to 63 mV/dec [65], which is close to the theoretical limit of $SS = 2.3k_BT/q \sim 60$ mV/dec at room temperature for a FET [66]; here, k_B is the Boltzmann constant, T is temperature in °K, and q is the electron charge.) Hole effective mobility ($\mu_{eff,h}$) of s-SWNTs ranges from ~10^3–10^5 cm^2/V-sec [5,6,25,63,67] and increases with d [5]. Such $\mu_{eff,h}$ value depends on CNT growth conditions and is better for CVD-grown CNTs compared to solution-processed CNTs [28].

In contrary to s-SWNTs, FETs with single metallic (m-) SWCNT-FETs have very small gate modulation and show ambipolar behavior with a smaller n-type tail (Figure 5a). Such gate modulation is often attributed to the presence of Mott-insulating states [68] and/or strain-induced bandgap [69] in m-SWCNTs. FETs with s-SWCNTs also show ambipolar behavior, more significantly at higher $|V_{DS}|$ and for larger diameter s-SWCNTs (Figure 5b) due to band-to-band tunneling (BTBT) [70].

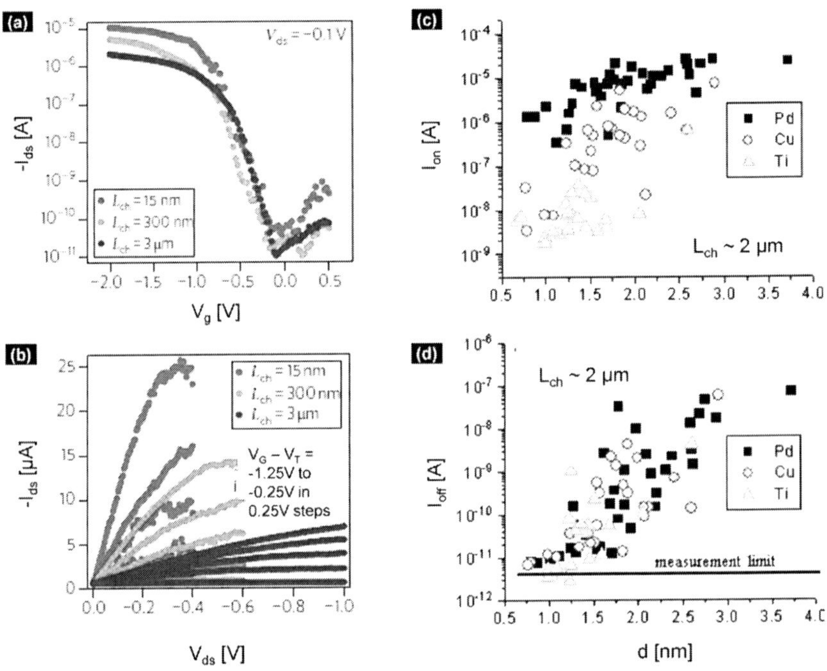

Figure 4. (**a**) Transfer or I_{ds}-V_g and (**b**) output or I_{ds}-V_{ds} characteristics of single s-SWCNT-FETs (reprinted by permission from Macmillan Publishers Ltd: Nature Nanotechnology [63], © 2010). Diameter dependence of (**c**) on-current I_{on} and (**d**) off-current I_{off} for s-SWCNT-FETs using different types of metals (palladium: Pd, copper: Cu, titanium: Ti) as source/drain (reprinted with permission from Ref. [58], © 2006 American Chemical Society.). In general, easier hole conduction for higher work-function (Φ_{WF}) contact metals like Pd results higher I_{on} and I_{off} for associated FETs.

2.4. FETs with Multiple SWCNTs

Multiple SWCNTs in FET's channel are either arranged randomly as network or aligned parallel to each other bridging source and drain. Network SWCNT-FETs can be made using CVD-grown SWCNTs with mixed chiralities (Figure 6a shows a scanning electron microscope (SEM) image for one such bottom-gated FET near a source/drain contact). These FETs have low on/off ratio, unless the conduction through metallic SWCNTs is turned off via periodic stripping [9]. An alternate way of improving on/off ratio is to use solution-processed, chirality sorted s-SWCNTs for making FETs. However, such FETs show lower conductance compared to CVD-grown SWCNTs because of the presence of structural defects [28].

Network SWCNTs have impedances at the junctions of different SWCNTs; consequently, they have lower conductance and are suitable only for applications like non-volatile memories [14,15]. Parallel arrangement of aligned SWCNTs resolves the issue with junction impedance and promises applications of SWCNTs in high performance radio frequency (RF) electronics [47,71,72], and transparent electronics [73,74]. Figure 6b shows a SEM image of one such top-gated FET before gate dielectric and gate metal formation. These aligned array SWCNT-FETs are either fabricated by growing SWCNTs via CVD on special types of quartz or sapphire substrates, whose crystal orientations [49] and/or step-edges [75,76] help parallel alignment (with less than 0.1% imperfection [77]) of d ~ 0.7~3 nm SWCNTs [26,44,45], or fabricated by aligning solution-processed SWCNTs via dielectrophoresis [46]. CVD-grown SWCNTs are long (with a length of several hundreds of μm [77]) and have high conductivity [5,25,63,67], however, require the use of post-processing techniques like thermo-lithography [54] and laser irradiation [78] for removal of m-SWNTs, or electrical breakdown [79] for breaking the continuity of m-SWNTs, and hence, cause FETs to use the remaining s-SWNTs.

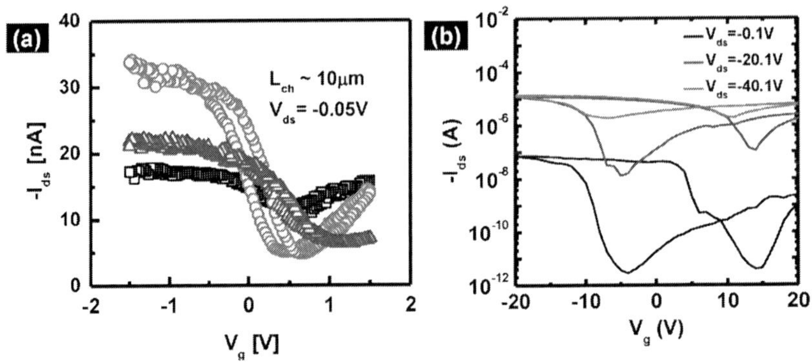

Figure 5. (**a**) Transfer characteristics of single metallic (m-) SWCNT-FETs showing ambipolar behavior (reprinted with permission from Ref. [44]; © 2012, AIP Publishing LLC). (**b**) Ambipolar behavior is also observed for FETs with a d ~ 1.7 nm s-SWCNT and at larger V_{ds} (reprinted with permission from Macmillan Publishers Ltd: Nature Nanotechnology [54], © 2013).

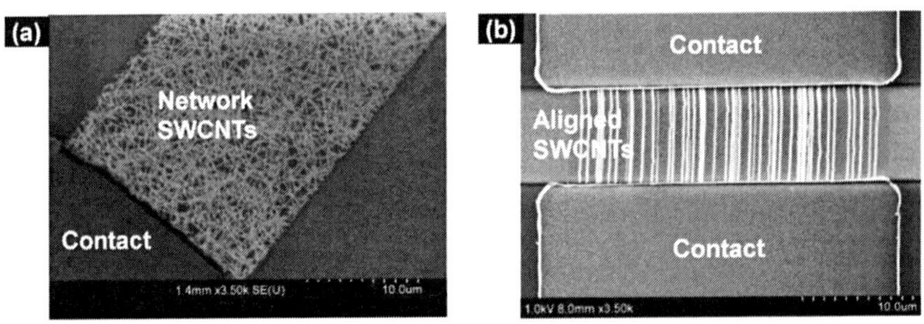

Figure 6. Scanning electron micrograph (SEM) of (**a**) random network of SWCNTs near one of the source/drain contact and (**b**) horizontally aligned SWCNTs bridging source/drain contacts.

3. VARIABILITY IN SWCNT-FETS

3.1. Variability in FETs with Single SWCNT

Performance variation in FETs with single s-SWCNT arises from the variation in d, contact and gate metals' Φ_{WF}, L_{ch}, T_{ox} (oxide thickness), and interface and oxide defects from one FET to another. For FETs with L_{ch}>μm, variation in d can give rise to several orders of magnitude change in I_{ds}, as shown in Figure 7a for small $V_{ds} = -50$ mV. One can simulate such transfer characteristics for different d (dashed lines in Figure 7a) by considering the conductance of the FET (G_{ds}) as a series combination of conductances of s-SWCNT (G_{ss}; see Ref. [44] for appropriate expressions) and contact ($G_c = G_{c0}T_c$; where T_c is the transmission probability of carriers near the source/drain contact) with components from both electron (e) and hole (h), *i.e.*,

$$G_{ds} = G_{ds,e} + G_{ds,h} = \left(G_{ss,e}^{-1} + G_{c,e}^{-1}\right)^{-1} + \left(G_{ss,h}^{-1} + G_{c,h}^{-1}\right)^{-1}, \qquad (2)$$

and hence, calculate I_{ds} using $I_{ds} = G_{ds} \times V_{ds}$. Simulated results for 0.6 nm $< d <$ 1.75 nm are consistent with measurement at $V_g < V_t$, where G_{ss} dominates G_{ds}. The differences in the region of $V_g > V_t$, where G_c dominates G_{ds}, partly reflect variation in the fixed component of G_c [80] that is not considered in simulation and also reflect I_{ds} measurement limitation set at ~pA. Plots of diameter dependence of I_{on} ($\equiv I_{ds}$ @ $V_g - V_t = -1$V, $V_{ds} = 50$ mV), maximum transconductance $G_{m,max}$ ($\equiv$max $|\partial I_{ds}/\partial V_g|$), and V_t (Figure 7b) provide additional detail about the origin of performance variation in s-SWCNT FET. At large diameters, when the transmission of carriers through the Schottky barrier near the contacts is unity, variation of $I_{on} \sim d$ and $G_{m,max} \sim d^2$ originates from the diameter dependence of μ_{eff} in s-SWNTs [5,81,82]. Similarly, variation of V_t ($\equiv V_g$ @ $|I_{ds}| = |I_{ds,max}|/100$) with d follows that of I_{on}. At small diameters, non-linear dependence of transmission through the contact Schottky barrier leads to a non-linear variation in I_{on}, $G_{m,max}$, and V_t with d.

Simulated I_{on}, $G_{m,max}$, and V_t distributions (calculated using diameter distribution of Figure 7c and variations of I_{on}, $G_{m,max}$, and V_t *vs.* d from Figure 7b) suggest good consistency with respective measurements (Figure 7d–f). Similar to the diameter distribution of Figure 7c, I_{on} distribution fits log-normal statistics, except near the lower tails of the distribution where transport through the contacts' Schottky barrier with $T_c < 1$ dominates I_{ds}. Measured $V_t - \langle V_t \rangle$ distribution (where $\langle V_t \rangle$ is the average of the distribution) is wider than the simulated one due to the extra contributions from defects

[16,83] and gate Φ_{WF} [84,85] variations that is not considered in the simulation. Passivation of SWCNT interface using hydrophobic self-assembled monolayers (SAMs) can tighten the V_t distribution by reducing hydroxyl (–OH) group related interface defects [86]. Figure 8a demonstrates utility of such SAM-based approach for reducing V_t distribution, where $L_{ch} << \mu m$ s-SWCNT FETs are made using solution-processed SWCNTs and passivated with hexamethyldisilazane (HMDS). As in CVD-grown s-SWCNTs, distribution in G_m (Figure 8b) for such solution-processed s-SWCNT FETs follows that of diameter (Figure 8c). In addition, contact resistance $R_c = 1/G_c$ in such small L_{ch} s-SWCNT FETs plays a dominant role in transport. Distribution of R_c (Figure 8d) arises from variations in T_c with d [28] and also from variations in the contact's Φ_{WF}. In contrary, FETs with m-SWCNT and similar L_{ch} have tighter R_c distribution [28]; however, they still suffer from variations in I_{ds} (Figure 8e) and, hence, G_m (Figure 8f). Since m-SWCNTs are unattractive for high performance electronics applications requiring high I_{on}/I_{off}, parametric variations in m-SWCNT FETs are not well studied.

3.2. Variability in FETs with Multiple SWCNTs

Similar to single SWCNT-FETs, performance variation in FETs with multiple SWCNTs originates from variations in d, Φ_{WF}, L_{ch}, T_{ox}, and defects [87,88,89]. A simple translation of single SWCNT-FET analysis to multiple SWCNT-FET; however, they cannot explain related variations, mostly because of the added complexity coming from variations in the arrangements and types of multiple SWCNTs bridging source/drain contacts. Multiple SWCNT FETs can have random network or aligned SWCNTs with varied density of m- and s-SWCNTs, a wide range of variations in the orientation of SWCNT and its crossings (especially for network SWCNT FET), and a wide distribution in d. This section discusses performance variation in aligned array SWCNT-FETs that have significant promise for high performance device applications. Such FETs have negligible variation in SWCNT's orientation [77] and also have negligible SWCNT crossings (Figure 6b). For a discussion on variabilities coming from SWCNT's orientation and crossings, which are more relevant for network SWCNT-FETs, please refer to Ref. [41].

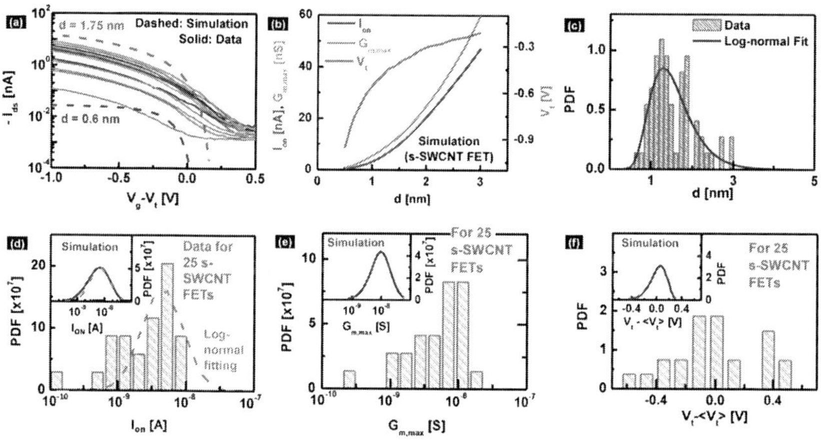

Figure 7. (**a**) Measured I_{ds} *vs.* $V_g - V_t$ ($V_t \equiv V_g$ @ $|I_{ds}| = |I_{ds,max}|/100$) characteristics for single s-SWCNT FETs (L_{ch} ~ 10 µm, V_{ds} = −50 mV) is within the simulated results for d = 0.6 nm and d = 1.75nm FETs. (**b**) Simulated I_{on} ($\equiv|I_{ds}|$@ $V_g - V_t$ = −1 V, V_{ds} = −50 mV), $G_{m,max}$ ($\equiv$max($\partial I_{ds}/\partial V_g$)), and V_t *vs.* d for s-SWCNT FETs. (**c**) Diameter distribution that is used to simulate performance distributions of s-SWCNT FETs in (**d–f**). Measured distributions of (**d**) I_{on}, (**e**) $G_{m,max}$, and (**f**) V_t for s-SWCNT FETs agree well with the simulated distributions (insets). For (**f**), <V_t> is the average of $V_t$distribution. These figures are reprinted with permission from Ref. [44]; © 2012, AIP Publishing LLC.

Solution-processed SWCNTs that are widely used for fabrication of FETs with network and partially aligned SWCNTs have yet to reach the target of 100% chiral selectivity. Some of the resultant FETs, therefore, have high on/off ratios when SWCNT density within the FET is low [35,46]. On the other hand, high performance electronics demand the availability of CVD-grown SWCNTs that always come with a mixture of m- and s-SWCNTs and require the application of post-

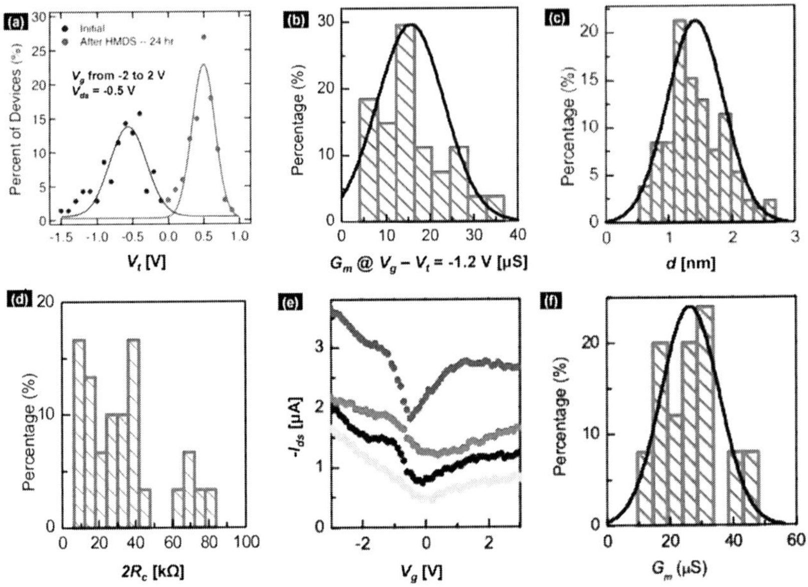

Figure 8. (**a**) Treatment of hexamethyldisilazane (HMDS) for 24 h narrows the V_t distribution significantly. V_g was swept from −2 V to 2 V for p-type, solution-processed s-SWCNT FETs during this measurement. Measured (**b**)G_m distribution and (**c**) diameter distribution for similar solution-processed, s-SWCNT FETs that has small channel length (L_{ch} = 150 nm). (**d**) Extracted contact resistance (R_c) distribution for L_{ch} = 150 nm, solution-processed s-SWCNT FETs. (**e**) Transfer characteristics and (**f**) G_m distribution for solution-processed, m-SWCNT FETs. (**a–c**) and (**d–f**) are reprinted with permission from Refs. [28] and [86], respectively; © 2012 American Chemical Society.

processing techniques (e.g., electrical breakdown [79], gas-phase reaction [90], thermo-lithography [54]) for having high on/off ratio in multiple SWCNT FETs. In some cases, application of post-processing techniques like electrical breakdown [79] and gas-phase reaction [90] also leads to unwanted removal of s-SWCNTs and, hence, performance degradation or even transistor failure [91]. Considering p_m as the probability of a single SWCNT in the FET being metallic, p_{Rs} as the probability of s-SWCNT being removed via post-processing, *IDC* as the index of dispersion in SWCNT count within the

FET, one can simulate the failure probability (p_F) for FETs with aligned SWCNTs at different SWCNT densities (Figure 9a). Such analysis, however, ignores a well-known aspect of 'inferential statistics' [92], *i.e.*, the distributions of SWCNT density and d in macro-scale (population distributions) is not always same as the distributions in micro-scale (sample distributions). As such, performance variability from diameter variations were expected to diminish for high density multiple SWCNT FETs via statistical averaging [25,41,91], making SWCNT density and m-SWCNT's presence as the major contributors to performance variations (Figure 9b) [91].

We performed a comprehensive experimental and theoretical study on FETs with aligned array SWCNTs [44] and ruled out the presence of statistical averaging in aligned-array SWCNT-FETs. Figure 10a plots distribution of I_{on} (measured at a $V_g - V_t = -1$ V) for aligned array-SWCNT FETs having $<N> \sim 11$ SWCNTs, where $<N> = <\rho>W$ is nominal number of SWCNT within the FET, W is the channel width, and $<\rho>$ is the average SWCNT density on the substrate per μm across the length of the nanotubes. Since variation in I_{on} follows Poisson statistics [44], we account for changes in μ_{Ion} (that arises from variation in sample preparation) from one set of array-SWCNT FET to another by dividing the standard deviation of $I_{on}/<N>$ (σ_{Ion}) with $\sqrt{\mu_{Ion}}$ (where μ_{Ion} is the average of $I_{on}/<N>$). Calculated $\sigma_{Ion}/\sqrt{\mu_{Ion}}$, normalized with respect to the same value measured for FETs with single SWCNT, shows a small reduction with the increase in $<N>$ (Figure 10b). If the diameter and density distributions of SWCNT for each array-SWCNT FETs (sample distribution) were the same as the wafer-level distribution (population distribution) of these parameters, as per central limit theorem [92], the normalized standard deviation should have reduced as $1/\sqrt{<N>}$ due to statistical averaging. The existence of deviation from $1/\sqrt{<N>}$, therefore, suggests significant variations in SWCNTs' density and diameter across the wafer, as was confirmed via extensive atomic force microscopy (AFM) at different locations over a macroscopic area of ST-cut quartz substrate that had CVD-grown aligned arrays of SWCNTs [44].

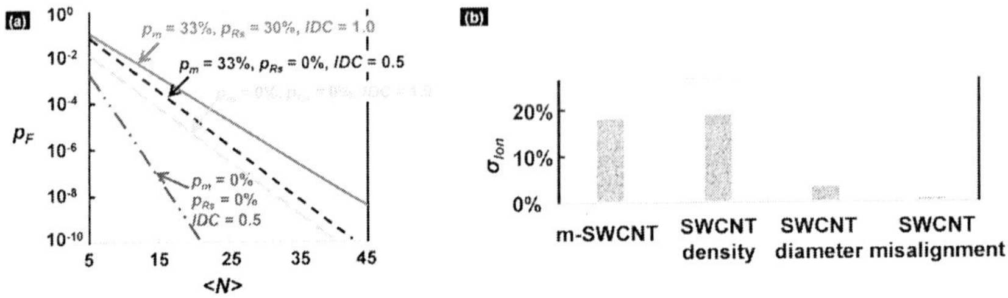

Figure 9. (**a**) Variation in failure probability (p_F; defined as probability of having no s-SWCNT within the FET) of m-SWCNT removed (via post-processing) aligned-array SWCNT-FETs with <*N*> (defined as the average number of SWCNTs within the FET before post-processing). Simulation is performed for different values of p_m (defined as the probability of having m-SWCNT within the FET before post-processing), p_{Rs} (defined as the probability of s-SWCNT removal via post-processing), and *IDC* (index of dispersion in SWCNT count). (**b**) Different variation sources like the presence of m-SWCNT before post-processing, SWCNTs' density, diameter, and misalignment contributes differently to the overall variation in standard deviation of I_{on} (σ_{Ion}). Simulation suggests that the presence of m-SWCNT and density distribution are the key variability sources in SWCNT FETs. These figures are plotted using data from Ref. [91].

To clarify the differences between measured $\sigma_{Ion}/\sqrt{\mu_{Ion}}$ and expected $1/\sqrt{\langle N\rangle}$ dependency, measured diameter and density variations across the wafer are used first to calibrate the variations measured at the microscopic level (*i.e.*, electronic properties of single-SWNT-FETs, as discussed in section 3.1). Later, the rules of "inferential statistics" [92] is used to analyze variations at the macroscopic level (*i.e.*, for array SWCNT FETs). Here, density variation is eliminated by counting the number of SWCNTs (*N*) for each array-SWNT FET and then using $(I_{ds}/N) - V_g$ characteristics for standard deviation calculation. Even after accounting for density variation, calculated σ_{Ion} at different <*N*> (normalized to its value for <*N*> = 1) shows significant deviation from $1/\sqrt{\langle N\rangle}$ scaling (Figure 10c). Such deviation suggests

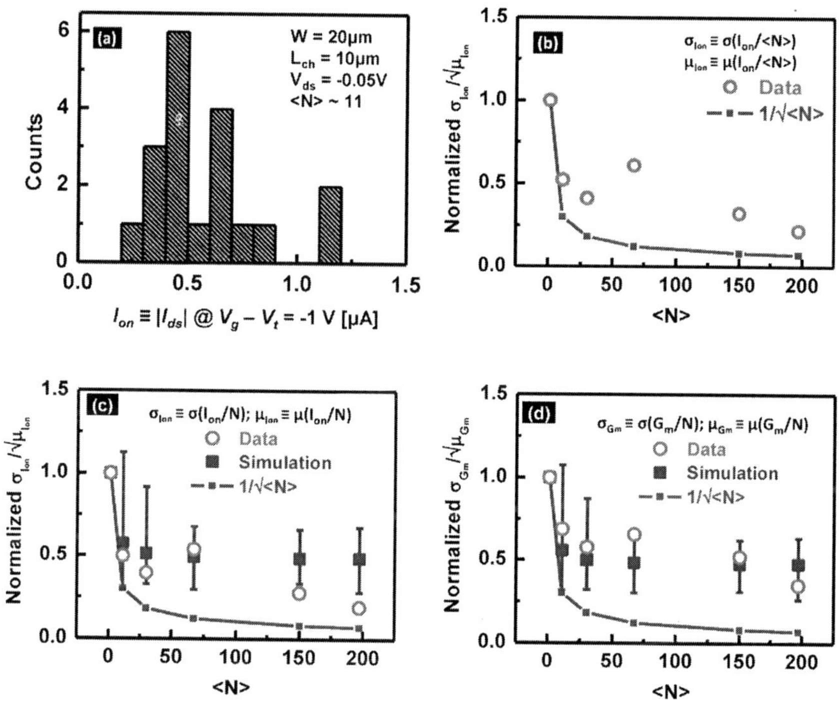

Figure 10. (**a**) Distribution of I_{on} among array-SWCNT FETs with mean SWCNT density $<N>$ ~ 11. (**b**) Normalized standard deviation of I_{on} (σ_{Ion}) *vs.* $<N>$ deviates from $1/\sqrt{<N>}$ scaling or central limit theorem. Normalized (**c**)σ_{Ion} and (**d**) σ_{Gm} for array-SWCNT FETs, where the effect of SWCNT density variation is removed by counting N for each FET and then calculating the standard deviations using I_{ds}/N–V_g characteristics of each FET. Variations inσ_{Ion} and σ_{Gm} still differs from $1/\sqrt{<N>}$ and required the consideration of wafer-scale diameter distribution for matching experimental trends via detailed simulation. Figures are reprinted with permission from Ref. [44]; © 2012, AIP Publishing LLC.

SWCNT density variation as a minor contributor to performance variation, thus contradicting the results of Refs. [25,41,91]. Finally, I_{ds}-V_g characteristics of array-SWCNT FETs are simulated for different $<N>$ by considering experimentally calibrated diameter dependence of single-SWCNT FETs's I_{ds}-$V_g$characteristics (Figure 7a) and measured (wafer-scale) density, diameter distributions of

SWCNT. Figure 10c,d show normalized standard deviations of I_{on} (σ_{Ion}) and $G_{m,max}$ (σ_{Gm}), as simulated for different <*N*> and their comparison to measured quantities. Since the sample sizes in measuring the standard deviations were small (~17–35), simulated standard deviations show a range of magnitudes for similar sample sizes within which the measured data points fits in. (The simulation framework [44], used for calculating σ_{Ion} and σ_{Gm} of array-SWCNT FETs, neglects contributions from m-SWCNTs. However, the procedure is suitable for demonstrating the importance of wafer-level variations in diameter and density).

Recently, use of smaller equivalent oxide thickness ($EOT \equiv \varepsilon_{SiO2}/\varepsilon_{HK} T_{ox}$; where ε_{SiO2}, ε_{HK} are dielectric constants of SiO_2 and high-κ dielectrics, respectively) has been shown to reduce the performance variation in FETs with single s-SWCNTs [93]. Decreasing *EOT* reduces the width of the Schottky barrier near the SWCNT and source/drain junction [57] and removes the long negative tails in the I_{on} distribution [44]. Moreover, as L_{ch}approaches the carrier mean-free path of SWCNT, I_{ds} saturates for large diameter SWCNTs and hence diameter dependence of I_{ds} is less pronounced [44]. Both these effects are studied by simulating small-scale array-SWCNT-FETs with *EOT* ~ 1 nm and L_{ch} = 300nm. (Here, simulated I_{ds}-V_g [44] is calibrated with measurements of Ref. [63]). Simulation suggests a decrease in σ_{Ion} for FETs with <*N*> = 1 with decreasing *EOT* (Figure 11a). However, at larger <*N*>, neither *EOT* nor L_{ch} scaling could improve the statistics because the effects of variations in density and diameter remain significant (Figure 11b). Therefore, narrowing diameter and density distributions are identified as the main areas for improvement via advanced growth and/or purification techniques to reduce performance variation in multiple-SWCNT FETs.

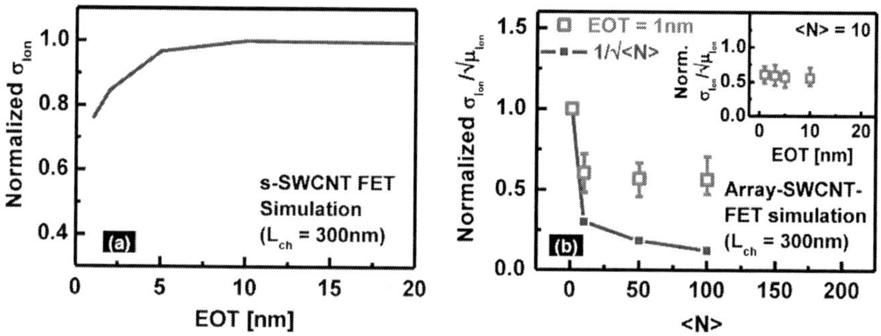

Figure 11. (**a**) Oxide scaling of s-SWCNT FET reduces σ_{Ion} (normalized by its value at equivalent oxide thickness (*EOT* = 20nm), as observed in Ref. [93]. (**b**) Normalized σ_{Ion} (*i.e.*, $\sigma_{Ion}/\sigma_{Ion,<N>=1}$) differs from $1/\sqrt{<N>}$ scaling for *EOT*= 1 nm array-SWNT FETs and show negligible effect of oxide scaling at fixed<*N*> (inset). Figures are reprinted with permission from Ref. [44]; © 2012, AIP Publishing LLC.

4. RELIABILITY OF SWCNT-FETS

Reliability (*i.e.*, device performance degradation) for any device can be interpreted by understanding its origin. For SWCNT-FETs, degradation arises from defects at the interface of SWCNT/oxide and also from defects within the oxide. Trapping and detrapping of carriers (*i.e.*, electrons and holes) into and out of these defects changes performance parameters like V_t, I_{ds}, G_m, *etc.* over time [94,95,96] and also causes hysteresis in the measured I_{ds}-$V_g$characteristics [83]. In addition to these interface and oxide defects, performance of SWCNT-FET is also affected by topological or Stone-Wales defects [97,98]. They affect the transport of carriers through SWCNT; hence, they lower the mobility of associated FETs. On the other hand, controlled introduction of defects via UV/ozone/chemical treatment [99], doping [20,100], *etc.* enable the application of SWCNTs as sensors by providing low-energy adsorption sites for introduction of chemical species. These structural defects, however, do not evolve with time during the operation of SWCNT-FET, hence do not affect reliability.

In this section, we discuss the nature of interface (section 4.1) and oxide (section 4.2) defects in SWCNT-FETs and show how these defects raise reliability concerns such as hysteresis (section 4.3), low-frequency noise (section 4.4), radiation damage (section 4.5), and time-dependent performance degradation (section 4.6). These interface and oxide defects though cause reliability concerns at different *quantities* depending on the orientation and density of SWCNTs within the FET, the *qualitative* features of reliability remain the same for any orientation and density, as studied in context of hysteresis and performance degradation in Ref. [94]. Therefore, in contrary toSection 3, where variability of FETs with single and multiple SWCNTs is discussed separately, a general discussion of reliability without reference to SWCNTs' density and orientation within the FET is sufficient for this section.

4.1. Interface Defects in SWCNT-FETs

Hydroxyl group (–OH), commonly present at any oxide surface [101,102], is the source of interface defects in SWCNT-FET [83,103,104]. Physisorbed water molecules and oxygen, present at the SWCNT/oxide in ambient condition, control –OH density [102] through electrochemical reaction: $O_2 + 4H^+ + 4e^- \leftrightarrow 2H_2O$ [102,105]. These –OH groups or interface defects surround SWCNTs laying on SiO_2 surface (Figure 12a) and can give rise to large hysteresis in measured I_{ds}-V_g characteristics [83,103,104], as well as potential time-dependent degradation in measured current [95,96]. Application of positive V_{gs} negatively charges –OH groups via electron capture from SWCNT positively shifts V_t and enhances hole conduction in commonly used p-type SWCNT-FET. Application of negative V_g, on the other hand, discharges the –OH group into neutral state, negatively shifts V_t and reduces hole conduction. Marcus-Garischer theory is generally used for modeling charging into or discharging from interface defects in SWCNT-FETs [94,105].

High temperature annealing under vacuum can reduce –OH groups [83,104] by eliminating water molecules at temperatures >190 °C [101]. Annealing at even higher temperature (at >1000 °C [101]) in vacuum would have completely removed the –OH groups; however, they have never been tested because of the high potential for damage to the SWCNTs and to the characterization system at such high temperatures. In addition to annealing in vacuum, –OH group or interface defects can also be reduced by using poly-methyl methacrylate (PMMA) encapsulation with thermal annealing [83] or by growing SWCNTs on different types of hydrophobic surfaces like octadecyltrichlorosilanated-SiO_2 [103,106], hexamethyldisilazanized-SiO_2 [86], octachlorotrisiloxane-capped organic dielectrics [107], *etc.*

4.2. Oxide Defects in SWCNT-FETs

Defects that are present within the gate dielectric, either near the SWCNTs or deep into the dielectric, also affect device performance. These oxide defects are broken bonds (Figure 12b) and pre-exist within the amorphous oxide network at various quantities, depending on the type of oxidation being performed. These pre-existing oxide defects can be annealed, for example, at high temperatures in hydrogen ambient [108]. Oxide defects are also generated near the vicinity of SWCNT during its growth, unless appropriate measures are taken to reduce the defects by running the growth in an inert environment [94]. In addition to the pre-existing ones, oxide defects are also generated during operation of FETs and causes dielectric breakdown [109,110,111]. The nature and origin of these oxide defects have been studied extensively using electron paramagnetic resonance [112,113], stress-induced leakage current [109], sweep-rate dependent I_{ds}-V_g hysteresis [94], and optical excitation [114]. Unlike interface defects in SWCNT-FET, oxide defects are ambipolar and can either switch between negative and neutral charged states (*i.e.*, act as electron traps) or switch between positive and neutral charged states (*i.e.*, act as hole traps) [115]—thus, showing a wide range of reliability concerns as discussed below.

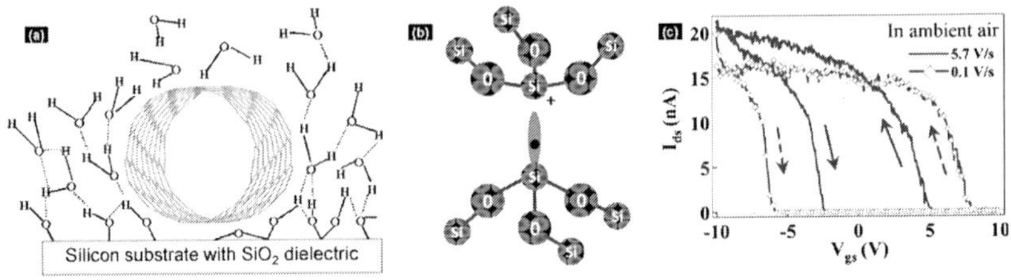

Figure 12. (**a**) Schematic representation of hydroxyl (–OH) group present at the surface of silicon dioxide on which SWCNTs are grown or placed. These –OH groups act as interface defect in SWCNT FETs. (**b**) Schematic representation of oxygen-vacancy related oxide defects [113] within silicon dioxide. (**c**) Transfer characteristics of s-SWCNT FET show large hysteresis between V_g sweeps performed in positive to negative and negative to positive directions. Measured hysteresis depends on sweep rate of V_g. (**a**,**c**) are reprinted with permission from Ref. [83]; © 2003 American Chemical Society.

4.3. Hysteresis in SWCNT-FETs

Large hysteresis in I_{ds}-V_g characteristics is a common phenomenon in SWCNT-FET and its magnitude depends on the concentration of defects in the FET [83,94], density of SWCNT [116], and also on the magnitude [94,107,117] and sweep-rate of V_g [83,94] (Figure 12c). Two stable charged and discharged states for the defects and switching between these states by application of appropriate V_g is the origin of hysteresis. Though such hysteresis promises memory applications using SWCNT, it is detrimental for device applications.

As mentioned before, both interface and oxide defects can give rise to hysteresis in SWCNT-FET. According to Marcus-Garischer theory [118], –OH groups or interface defects in SWCNT have separate, broad energy levels for discharged (neutral) and negatively charged state, as shown in Figure 13a. In FETs where interface defects predominantly controls hysteresis, the separation between these neutral and charged energy levels of –OH (known as reorganization energy) controls the magnitude of hysteresis [94]. Charging and discharging of oxide defects, on the other hand, can be modeled within a Shockley-Read-Hall framework [119]. Density and type of oxide defects depend on the oxide quality on which carbon nanomaterials are grown or placed on [94]. In some harsh growth conditions, oxides near the FET channel are damaged and give rise to large amount of near-interface oxide defects that dominate the trends in measured hysteresis. In other optimized conditions, we only have a small amount of bulk oxide defects that controls the hysteresis trends.

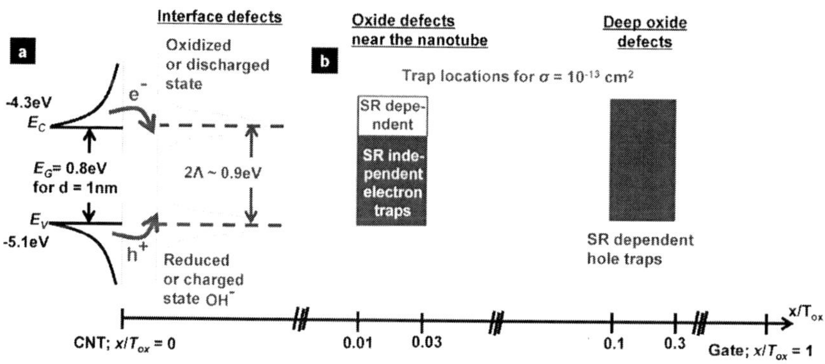

Figure 13. (**a**) Schematic representation of the Marcus-Gerischer theory [118] used to represent the dynamics of –OH group related interface defects to simulate hysteresis in a SWCNT FET with one d = 1 nm s-SWCNT [94]. E_C,E_V, E_G are the conduction band, valence band, bandgap of the s-SWCNT, and Λ is the reorganization energy [105]. (**b**) Representative spatial and energy locations of near-interface electron traps (*i.e.*, oxide defects that switch between negative and neutral charged states) and deep hole traps (*i.e.*, oxide defects that switches between neutral and positive charged states) with σ ~ 10^{-13} cm^2 [120] used for simulating oxide-defect related hysteresis in Ref. [94]; © 2012 WILEY-VCH Verlag GmbH & Co. KGaA, Weinheim.

Self-consistent Poisson and drift-diffusion solution [94], supported by detailed experiments, suggest contrasting sweep rate dependence of hysteresis for different types of defects in SWCNT-FET. Interface defects are characterized by short charge capture and emission times compared to the time used for sweeping V_g to measure hysteresis and, therefore, have sweep rate independent hysteresis. On the other hand, near-interface oxide defect have small capture but comparatively larger emission times, such that hysteresis increases with sweep rate; whereas oxide defects within the bulk of the dielectric have large capture and emission times, such that hysteresis decreases with sweep rate. Measurement of hysteresis at different sweep-rate dependencies, therefore, is used to identify the contributions from different types of defects on hysteresis in SWCNT-FET and modify the fabrication steps appropriately to optimize different defects. Growing SWCNTs in an inert environment reduces the amount of near-interface oxide defects [94] and enables one to employ different interface defect removal techniques to compare their effectiveness. A comparison of different –OH group reduction methods (e.g., heating FETs in vacuum [83], coating SWCNT with PMMA [83] or spin-on-glass [94];Figure 14a) suggests high temperature annealing, quality of coatings to block water penetration, and reaction of OH-group with coated molecules as the key for reducing hysteresis originating from –OH group related interface defects. All these –OH removal techniques have shown negligible hysteresis when SWCNT-FET is swept with smaller $|V_{g,max}|$. Table 1 summarizes the values of $|V_{g,max}|$ and corresponding maximum of average oxide electric fields ($E_{ox,max}$) that has been reported to show negligible hysteresis under various conditions [9,83,94,103,106,107,121,122]. ($E_{ox,max}$ calculation steps for all the references are shown in the caption of Table 1.) Among these conditions, SWCNT-FETs with spin-on-glass/HfO_2 gate dielectric shows negligible hysteresis for highest $E_{ox,max}$ ~ 0.45 MV/cm. At higher $|V_{g,max}|$ or $E_{ox,max}$, hysteresis reappears [94,107] which suggests incomplete removal of interface and oxide defects using the techniques mentioned above. Use of combinatorial approaches such as placing SWCNT on hydrophobic surfaces, annealing at temperatures above 400 °C in an inert environment, and then coating with hydrophobic gate dielectric may eliminate hysteresis even at higher $E_{ox,max}$. Hysteresis reduction is critical for SWCNT-FET, because reduced hysteresis is generally accompanied with a narrower distribution in V_t (Figure 8a) and also for the hysteresis itself (Figure 14b) [86]. Removal of –OH group from SWCNT-FETs, therefore, is necessary to improve variability in SWCNT-FET.

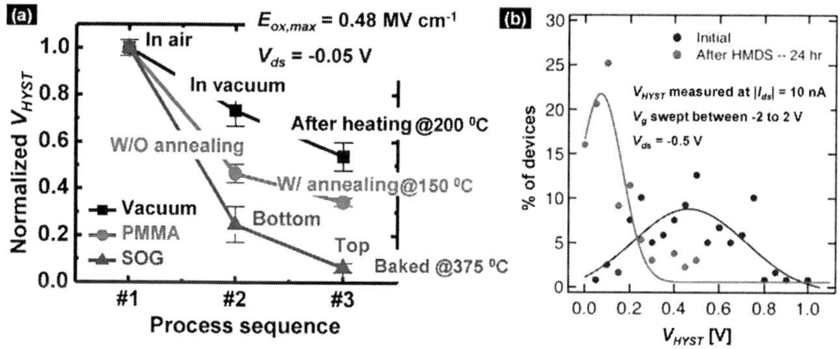

Figure 14. (**a**) Comparison of effectiveness of different hysteresis reduction techniques like heating in vacuum, use of poly-methyl methacrylate (PMMA) and SOG. Measured hysteresis (V_{HYST}) in each case at similar$E_{ox,max}$ is normalized with that measured in air. For heating in vacuum, hysteresis is first measured in air at room temperature, then after 10 h in vacuum (10^{-5} Torr), and finally after 10 h of heating at 200 °C followed by cooling to room temperature in vacuum. For PMMA passivation (with: W/ and without: W/O), hysteresis is first measured in air before PMMA coating, then after PMMA coating and heating for 1min at 110 °C, and finally after an additional 15 h annealing at 150°C. For SOG encapsulation, hysteresis is first measured in air using a bottom-gated configuration, then after SOG encapsulation and 375 °C baking using bottom- and top-gated configurations. This figure is taken from Ref. [94]; © 2012 WILEY-VCH Verlag GmbH & Co. KGaA, Weinheim. (**b**) Use of hysteresis reduction technique like HMDS treatment before placing SWCNTs on oxides is also effective for narrowing distribution of measured hysteresis. This figure is reprinted with permission from Ref. [86]; © 2012 American Chemical Society.

4.4. Low-Frequency Noise in SWCNT-FETs

Low-frequency noise measurement is a general way of characterizing deformations or defects in any nanoscale systems [123,124,125,126,127]. It is also a reliability concern for flash memories [128,129], radio-frequency (RF) circuits [130], and, in general, for many different digital and analog applications employing nanoscale FETs. Among different components of low-frequency noise, the following components are routinely measured: (i) the frequency independent thermal noise component with power spectral density of $4k_BTR$, which originates from the statistical Brownian motion of carriers at temperature T in a resistive system with resistance R; (ii) the Lorenzian-type noise component with power spectral density ~ $1/[1 + (2\pi f\tau)^2]$, which originates from trapping/detrapping events for a small number of deformations or defects with time constant τ and cut-off frequency $f_c = 1/\tau$; and (iii) the flicker noise component with power spectral density ~ $1/f^n$, which originates from a distribution of trapping/detrapping events.

Noise in SWCNT-FETs has been widely studied to understand the origin of deformations or defects in SWCNT lattice and in its surrounding oxides [123,124,125,128,129], and also to explore the effect of radiation damage [133] and chemical exposure [134]. Measured low-frequency noise in SWCNT is much more pronounced than in silicon electronics [135,136] because of the presence of a few carriers for conduction [134]. In addition to structural defects in the SWCNT lattice [135], oxide defects contribute significantly to low-frequency noise behaviors in SWCNT-FETs [137]. (A recent discussion on different sources of low-frequency noise in SWCNT materials and devices can be found in Ref. [138].) Interface defects interact at faster time-scales and hence noise because interface defects are generally overwhelmed by thermal noise at high frequencies. Oxide defects, on the other hand, have comparatively longer and distributed charging and discharging times (either due to distribution in tunneling distance [126,127] or due to distribution in defects' energetics [139]) that give rise to $1/f$ behavior in the power spectral density of measured noise in I_{ds} at low frequencies [134,135], as shown in Figure 15a, which is measured for a single s-SWCNT-FET at different V_g [134]. From Hooge's number fluctuation theory [140] noise power spectral density (S_I) can be expressed as,

Table 1. Values of *EOT*, SWCNT density (*D*), capacitance ratio (Ξ) [131], maximum electric field with parallel plate assumption (*i.e.*, SWCNT is presumed to form a dense film), and maximum of the average oxide electric field ($E_{ox,max}$) for reported instances [9,83,94,103,106,107,121,122] of negligible hysteresis in SWCNT-FETs. Here, $\Xi = 4\pi \times EOT/\varepsilon_{SiO2} \times \lambda_0[2/\varepsilon_{SiO2} \times ln[\lambda_0/d \times sinh(2\pi \times EOT/\lambda_0)/\pi] + C_Q^{-1}]$ [131] (where $\lambda_0 = 1000/D$ is the average distance between each SWCNT, $C_Q = 4 \times 10^{-12}$ F/cm is the quantum capacitance [132]), $E_{P,max} = V_{G,max}/EOT$, and $E_{ox,max} = \Xi \times E_{P,max}$.

Reference	EOT [nm]	D [μm^{-1}]	Ξ	$V_{g,max}$ [V]	$E_{P,max}$ [MV cm^{-1}]	$E_{ox,max}$ [MV cm^{-1}]	Type of Dielectric
[9] Cao *et al.*, Nature '08	8.2	~6	0.01	2	2.44	0.02	HfO_2
[106] Weitz *et al.*, NL '07	20.3	5.48	0.16	2	0.98	0.16	Octadecyltrichlorosilanated-SiO_2
[107] Hur *et al.*, JACS '05	6.9	1 (Single CNT)	0.01	1	1.45	0.02	Octachlorotrisiloxane-capped organic dielectric
[83] Kim *et al.*, NL'03	500	1 (Single CNT)	0.36	10	0.2	0.07	SiO_2 with PMMA coating
[103] McGill *et al.*, APL '06	100	~3	0.30	2	0.2	0.06	Octadecyltrichlorosilanated-SiO_2
[121] Hu *et al.*, Phys. E '08	200	~3	0.45	10	0.5	0.23	Aminopropyltriethoxysilaned-SiO_2
[131] Kim *et al.*, APL '07	6.5	1 (Single CNT)	0.01	1	1.54	0.02	Al_2O_3
[94] Jin *et al.*, AFM '12	40	~6	0.29	6	1.5	0.45	Spin-on-glass (SOG) + HfO_2

$$S_I = I_{ds}^2 \frac{A}{f}, \text{ where } A \sim \frac{\alpha_H}{I_{ds}}. \quad (3)$$

where α_H is a fitting parameter that varies from device to device. Equation (3) suggests a reduction in noise amplitude *A* with the increase in carrier number or I_{ds}, as measured for the p-type FET for different V_g (Figure 15b) [134].

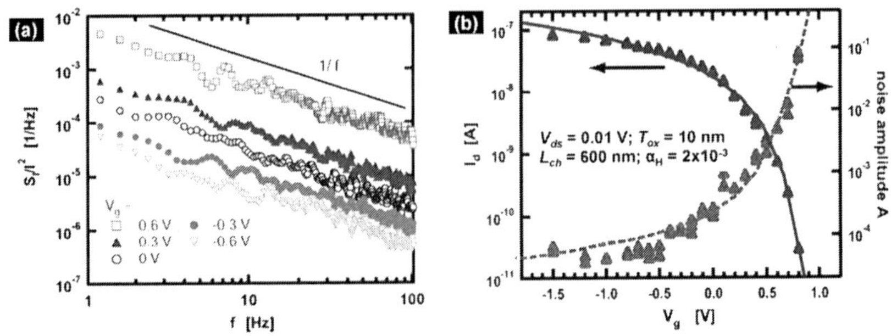

Figure 15. (**a**) Flicker or 1/*f* noise measured for single s-SWCNT FET at different V_g. (**b**) Increase in V_g decreases I_{ds} and hence increases noise magnitude following Hooge's number fluctuation theory [140]. These figures are reprinted with permission from Ref. [134]; © 2006 American Chemical Society.

Noise measurements in cryogenic conditions reduce thermal noise and also freeze-off some of the defects that contribute to noise at higher temperatures. Responses from the remaining few oxide defects result in random telegraph noise (RTN) with stochastic variation in I_{ds} and discrete current levels [137,141,142,143]. Figure 16a shows RTN obtained from a SWNT-FET (measured at T = 4.2 °K) suggesting four major switching levels resulting from a slow switching defect (whose fluctuation is digitized in the bottom of Figure 16a) and a fast switching defect [142]. Corresponding power density spectrum for these fluctuations is shown in Figure 16b that suggests two overlapping Lorenzian spectrums with corner frequencies (f_c) at 0.2 Hz (for the slow defect) and 5.8 Hz (for the fast defect). At higher temperatures (T = 60 °K), however, responses from these two defects are mixed up (inset of Figure 16b) and noise from new defects (with higher f_c) shows up. Similar noise analysis at different V_g(Figure 16c) enables one to extract distributions of capture time τ_c (defined as the time required for a high to low transition in I_{ds}) and emission time τ_e (defined as the time required for a low to high transition in I_{ds}). These distributions are used to calculate the averages of capture and emission times (<τ_c> and <τ_e>) at different V_g(Figure 16d) and, hence, the location of defect causing RTN from the slope of ln(<τ_e>/<τ_c>) *vs.* V_g plot [137].

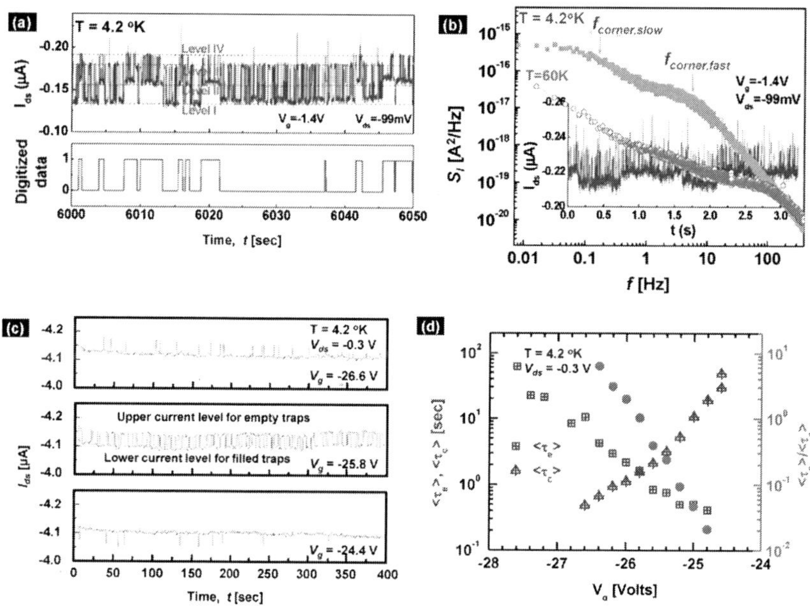

Figure 16. (**a**) Random telegraph noise (RTN) measured in a SWCNT-FET at 4.2 °K with V_g = −1.4 V and V_{ds} = −99 mV shows four major switching levels (red dashed lines) due to the presence of one slow (with larger trapping/detrapping time) and one fast (with smaller trapping/detrapping time) defect. The digitized fluctuation for the slow defect is shown in the bottom graph. (**b**) Noise power spectral density of Figure 16a at 4.2 °K and also for the same device at 60 °K. The inset shows the time domain current data at 60 °K. Spectral density at 4.2 °K shows the cut-off frequencies designating the presence of slow and fast traps. Response from these defects disappears at higher temperatures. (**c**) Low temperature RTN measurements in another SWCNT-FET at different V_g. (**d**) Voltage dependence of average capture <τ_c>, emission <τ_e> times and their ratios. (**a**,**b**) are reprinted with permission from Ref. [142]; © 2008 American Chemical Society and (**c**,**d**) are replotted using data from Ref. [137].

Though RTN analysis has not yet been employed to study the evolution of defects—*i.e.*, reliability in SWCNT-FETs—*1/f* noise has been used to study the radiation effects in such FETs [133]. An increase in 1/*f* noise after gamma radiation (Figure 17) suggests generation of defects within the oxide. The next section discusses the radiation effects on SWCNT-FETs in detail.

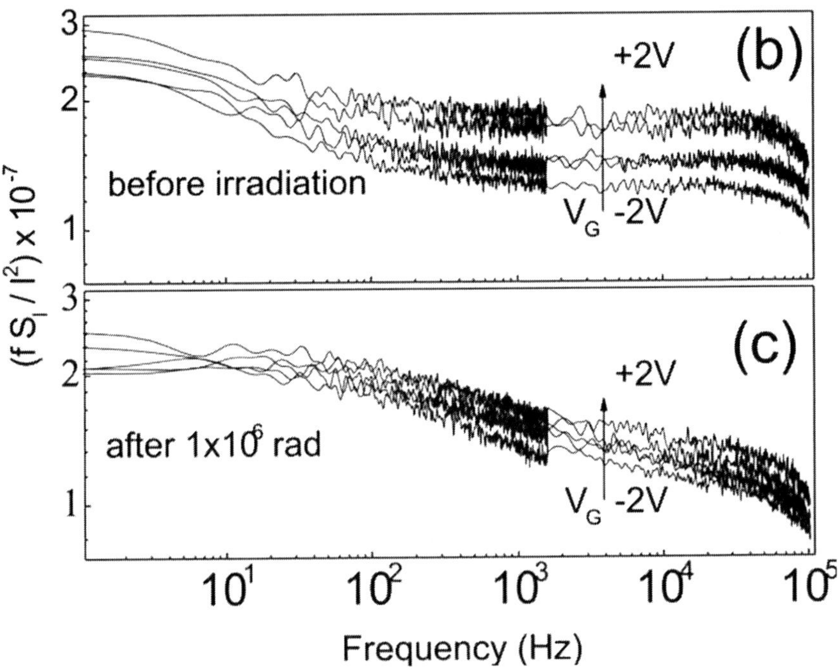

Figure 17. Measured noise power spectral density at different V_g before and after gamma radiation (reprinted with permission from Ref. [133]; © 2010, AIP Publishing LLC).

4.5. Radiation Damage in SWCNT-FETs

Promising performance parameters of SWCNTs have initiated studies to explore their potential in terrestrial and space-based applications [133,144,145,146,147,148,149]. In such niche applications, radiation tolerance is one of the key reliability parameters. In this regard, use of SWCNT as a channel showed significant advantages over its silicon-based predecessors. Small cross-sections of SWCNT make it less susceptible to incident radiation particles, unless radiation dose (termed as displacement damage dose, DDD) exceeds ~ 10^{13} MeV/g [150]. One can evaluate DDD for different types of particles (e.g., alpha, proton, ionized chemical species) by considering the number of particles striking SWCNTs per unit area (*i.e.*, fluence) and the efficiency of these particles in creating defects within the SWCNT lattice (which is quantified using the term non-ionizing energy loss [150]). Damage in SWCNT at high radiation dose (with DDD > 10^{13} MeV/g) can be monitored using Raman spectroscopy as an increase in intensity at the defect (D)-band wavenumber of ~ 1300 cm^{-1} [133,146,147,148,151] and also using electrical measurement as an increase in resistance [148,151]. Figure 18a,b show Raman spectra for m- and s-SWCNT-FETs measured at different amount of ionized boron irradiation. Radiation gradually introduces defects within the SWCNT lattice, therefore, increasing the ratio of Raman intensities measured for D- and G^+-bands or for D- and G'-bands. (G^+- Raman signal comes from the atomic vibration of carbon atoms along the axial direction of SWCNT and peaks at a wavenumber of ~1590 cm^{-1} with no diameter dependence. On the other hand, G'- Raman signal arises from a two-photon, inter-valley, second-order Raman scattering process and peaks at a wavenumber of ~2700 cm^{-1} with significant diameter dependence. Both G^+ and G' have no influence from structural defects in SWCNTs. (Dresselhaus *et. al* [152] provide an elaborate review of Raman signal and its origin in different types of SWCNTs.) Ratio of Raman intensities obtained using D- and G'-bands (D/G') are shown to be more sensitive in monitoring the effect of radiation that linearly depends on DDD (Figure 18c), when D/G' at a particular amount of radiation (Φ) is normalized to its value measured before radiation ($\Phi = 0$) [148]. The same linear degradation law also holds for the radiation-induced resistance degradation in SWCNT and graphene.

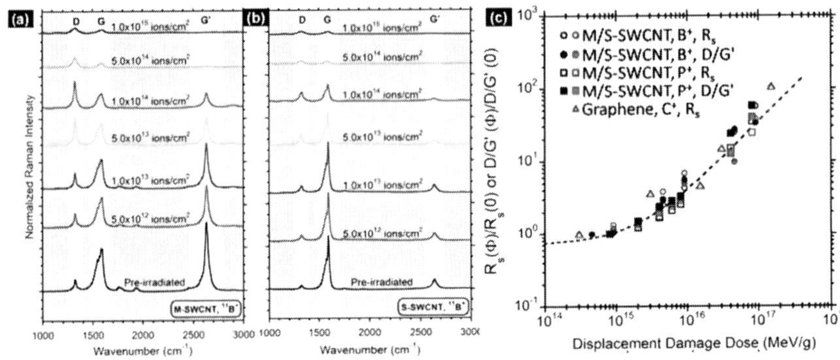

Figure 18. Raman spectra for (**a**) m-SWCNT and (**b**) s-SWCNT thin-films irradiated with 150 keV $^{11}B^{+}$ ion. The pre-irradiated spectrum is given as a reference and the intensity of each irradiated sample is normalized to the G band intensity of the reference. (**c**) Changes in the normalized (with respect to pre-radiated sample) Raman D/G′ ratio and normalized sheet resistance (R_S) with displacement damage dose (DDD) for boron-ion (B^+) and phosphorous ion (P^+) irradiated m-SWCNTs and s-SWCNTs, and carbon-ion (C^+) irradiated graphene (Ref. [153]). The dashed line is a linear fit of all D/G' ratioand R_S values. Figures are reprinted with permission from Ref. [148]; © 2012, AIP Publishing LLC.

At radiation doses comparable to aerospace environments (*i.e.*, for DDD<10^{13} MeV/g), only the dielectric layer of SWCNT-FETs are damaged with no damage to the SWCNT lattice [133,146,154]. Radiation in such condition changes V_t of SWCNT-FETs with negligible change in mobility (Figure 19a), therefore, suggests negligible defect formation within the SWCNT. The behavior is also confirmed via Raman spectroscopy (Figure 19b) that shows negligible change in D-band and the radial breathing mode (RBM) Raman signals. (RBM signal comes from the atomic vibration of carbon atoms along the radial of SWCNT and has peaks at wavenumbers in the range of ~100–400 cm^{-1} with strong diameter dependence [152].) A decrease in V_t with increasing radiation for the p-type SWCNT-FET (Figure 19a) suggests creation of oxide defects that act as hole trapping sites. The situation improves significantly when a comparatively thinner (~23 nm) silicon oxynitride dielectric (grown via plasma-enhanced chemical vapor deposition) replaces the regularly used ~100 nm SiO_2 in making SWCNT-FETs for radiation testing [145]. Pre- and post-radiation I_{ds}-V_g measurements (Figure 20) suggest negligible change in FET parameters above threshold (*i.e.*, V_t and μ_{eff}). Creation of balanced hole and electron trapping sites within the dielectric [155] and use of thin dielectric that allows the escape of trapped charges [156] are presumed to be the origin of such radiation hardness in these devices. The sub-threshold behavior, on the other hand, shows degradation when the devices are exposed to a negative gate electric field of −1 MV/cm. The origin of such degradation has been attributed to the formation of electron trapping sites within the oxide or at the interface of SWCNT—which requires detailed future analysis. Overall, SWCNT-FETs with thin dielectric offer great potential for space electronics in the future.

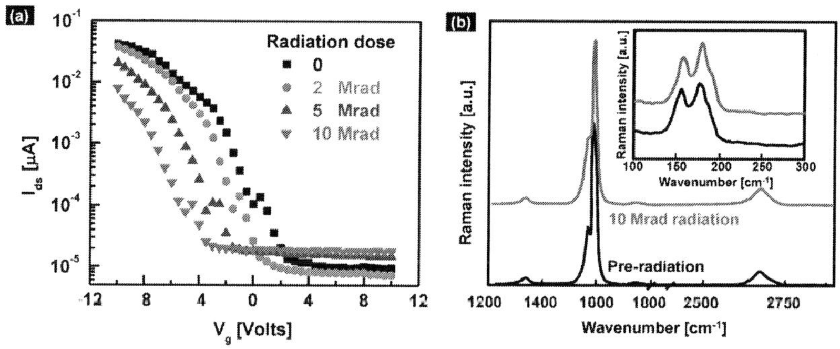

Figure 19. (**a**) Transfer characteristics of an SWCNT-FET measured in vacuum for different doses of Co-60 gamma rays. (**b**) G-band and G'-band Raman spectra of SWCNTs (98% s-SWCNT enriched thin film) within the FET before (pre-) and after (post-) irradiation in vacuum. Inset shows the Raman spectra in the low wavenumber regime that suggests negligible change in radial breathing modes. These figures are replotted using data from Ref. [146].

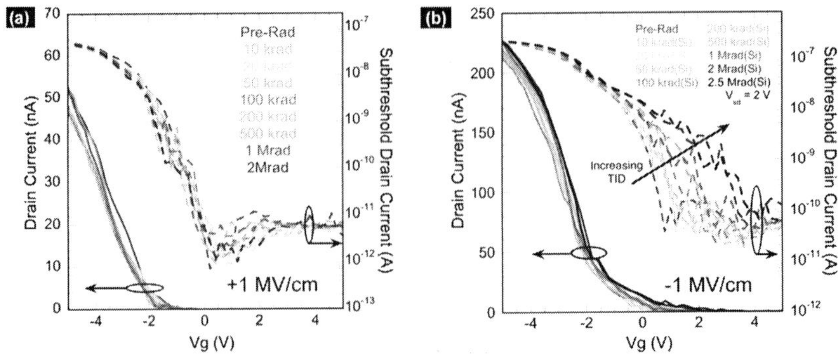

Figure 20. Transfer characteristics of an s-SWCNT FET fabricated with radiation-hard silicon oxynitride gate dielectric. Irradiation is conducted under vacuum with (**a**) positive, and (**b**) negative gate electric field of 1 MV/cm. Figures are reprinted with permission from Ref. [145] © Cambridge University Press 2011.

4.6. Performance Degradation in SWCNT-FETs

Performance degradation of transistor parameters over time is a critical reliability concern that dictates the determination of device lifetime. In transistor reliability literature, bias temperature instability [157] and stress-induced leakage current [109] are often used to monitor performance degradation and to probe the dynamics of defects in different regions of the FET. Though hysteresis and low-frequency noise has been widely used to study the nature of defects in SWCNT-FET, detailed analysis of performance degradation over longer time scales is still lacking. Studies of degradation up to a few seconds [95,96] though provides critical information about the nature of some of the defects, projection of performance degradation up to an intended lifetime of 5–10 years requires accelerated tests (at voltages much higher than use condition) and for longer times (1000 s or more), which are widely used in conventional CMOS reliability studies [157].

Some of the recent long-term performance degradation studies in SWCNT-FETs [94] suggest oxide defect as the key contributor for the instability. SWCNT-FETs with a large amount of near-interface oxide defects show significant I_{ds} degradation (ΔI_{ds}; Figure 21a) and have a small time constant $\tau \sim 7 \times 10^4$ sec for the fitted stretched exponential forms according to $\Delta I_{ds} = I_{ds0}[1 - \exp(-t/\tau)^{\beta}]$ (where I_{ds0} is the magnitude of drain current at $t = 0$, t is the time at which degradation is measured, and τ, $\beta \sim 0.35$ are fitting parameters) [158]. On the other hand, devices with a small amount of bulk oxide defects have lower ΔI_{ds} and long $\tau \sim 2 \times 10^7$ s (with $\beta \sim 0.35$). Moreover, the nature of trapping into oxide defects remains the same at different accelerated stress conditions, as seen from the measured ΔI_{ds} at different gate biases (Figure 21b). All these degradation curves can be fitted with a fixed β and a voltage dependent τ (higher for higher V_g) that suggests more and deeper oxide defects are accessible at higher stress biases. Therefore, by using τ as a voltage dependent parameter, one can project reliability of these SWCNT-FETs at different use conditions. Detailed analysis of such degradation by understanding the trapping dynamics (which may require the use of one or multiple time constants [95]) is still lacking.

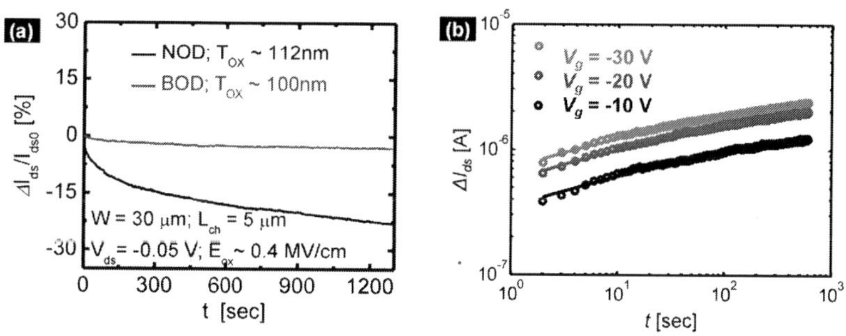

Figure 21. (**a**) Percentage change in I_{ds} (ΔI_{ds}) relative to its initial value (I_{ds0}) as a function of time (t) for two SWCNT-FETs, one having a large amount of near-interface oxide defects (NOD) and the other having a small amount of bulk oxide defects (BOD). The figure is taken from Ref. [94]; © 2012 WILEY-VCH Verlag GmbH & Co. KGaA, Weinheim. (**b**) Drain current degradation measured at different gate stress bias V_g. All the degradation curves can be fitted (lines) using a single expression: $\Delta I_{ds} = I_{ds0}[1 - \exp(-t/\tau)^{\beta}]$ with fixed β and voltage dependent τ (higher for higher V_g).

5. SUMMARY AND OUTLOOK

We have discussed different sources of variability and reliability that controls the change in time-zero and time-dependent performance parameters in SWCNT-FETs. Variability in SWCNT FETs performance is mainly attributed to the distributions of SWCNTs' diameter across the wafer and the presence of hydroxyl-group related interface defects in SWCNTs' surroundings. The removal of m-SWCNTs and the tightening of diameter distribution for s-SWCNTs via chiral-selective growth [159] or post-processing [50,54], and complete passivation of SWCNTs via removal of hydroxyl groups [83,94] are necessary for controlling variability. Removal of hydroxyl groups from SWCNTs' surfaces will also be the key to having a reliable SWCNT FET that will operate with reduced performance degradation over its lifetime. In addition, oxide defects keep playing a dominant role in controlling reliability parameters, like low-frequency noise, radiation damage, and long-term performance degradation. Therefore, solutions in terms of having reliable gate dielectrics with less defects [111,160] will not only help current CMOS scaling, but will also assist with the introduction of beyond CMOS devices with SWCNT-based FET.

ACKNOWLEDGMENTS

The author acknowledges contributions from John Rogers (University of Illinois) for introducing the presented area of research and also from Sunghun Jin, Frank Du (University of Illinois) for related experiments and discussion. Discussions with Muhammad A. Alam (Purdue University), Cory D. Cress (Naval Research Laboratory), Benji Maruyama (Air Force Research Laboratory), and funding from Air Force Office of Scientific Research (contract number: FA 8650-09-D-5037), Materials Structures and Devices Focus Center (Task ID: 2051.003.4) of Semiconductor Research Corporation are acknowledged for preparation of this review article.

REFERENCES

1. Iijima, S. Helical microtubules of graphitic carbon. *Nature* **1991**, *354*, 56–58.
2. Datta, S. *Quantum Transport: Atom to Transistor*, 2nd ed.; Cambridge University Press: Cambridge CB2 2RU, UK, 2005.
3. Saito, R.; Sato, K.; Oyama, Y.; Jiang, J.; Samsonidze, G.G.; Dresselhaus, G.; Dresselhaus, M.S. Cutting lines near the Fermi energy of single-wall carbon nanotubes. *Phys. Rev. B* **2005**, *72*.
4. Jorio, A.; Pimenta, M.A.; Souza, A.G.; Saito, R.; Dresselhaus, G.; Dresselhaus, M.S. Characterizing carbon nanotube samples with resonance Raman scattering. *New J. Phys.* **2003**, *5*.
5. Zhou, X.J.; Park, J.Y.; Huang, S.M.; Liu, J.; McEuen, P.L. Band structure, phonon scattering, and the performance limit of single-walled carbon nanotube transistors. *Phys. Rev. Lett.* **2005**, *95*.
6. Durkop, T.; Getty, S.A.; Cobas, E.; Fuhrer, M.S. Extraordinary mobility in semiconducting carbon nanotubes. *Nano Lett.* **2004**, *4*, 35–39.
7. Guo, J.; Goasguen, S.; Lundstrom, M.; Datta, S. Metal-insulator-semiconductor electrostatics of carbon nanotubes. *Appl. Phys. Lett.* **2002**, *81*, 1486–1488.
8. Rogers, J.A.; Someya, T.; Huang, Y. Materials and mechanics for stretchable electronics. *Science* **2010**, *327*, 1603–1607.
9. Cao, Q.; Kim, H.S.; Pimparkar, N.; Kulkarni, J.P.; Wang, C.J.; Shim, M.; Roy, K.; Alam, M.A.; Rogers, J.A. Medium-scale carbon nanotube thin-film integrated circuits on flexible plastic substrates. *Nature* **2008**, *454*, 495–500.
10. Sun, Y.; Rogers, J.A. Inorganic semiconductors for flexible electronics. *Adv. Mater.* **2007**, *19*, 897–1916.

11. Perelaer, J.; Smith, P.J.; Mager, D.; Soltman, D.; Volkman, S.K.; Subramanian, V.; Korvink, J.G.; Schubert, U.S. Printed electronics: The challenges involved in printing devices, interconnects, and contacts based on inorganic materials. *J. Mater.Chem.* **2010**, *20*, 8446–8453.

12. Cao, Q.; Hur, S.H.; Zhu, Z.T.; Sun, Y.G.; Wang, C.J.; Meitl, M.A.; Shim, M.; Rogers, J.A. Highly bendable, transparent thin-film transistors that use carbon-nanotube-based conductors and semiconductors with elastomeric dielectrics. *Adv. Mater.* **2006**, *18*, 304–309.

13. Xiong, F.; Liao, A.D.; Estrada, D.; Pop, E. Low-power switching of phase-change materials with carbon nanotube electrodes. *Science* **2011**, *332*, 568–570.

14. Kianian, S.; Rosendale, G.; Manning, M.; Hamilton, D.; Huang, X.M.H.; Robinson, K.; Kim, Y.W.; Rueckes, T. A 3D stackable Carbon Nanotube-based nonvolatile memory (NRAM). In Proceedings of the ESSDERC, Sevilla, Spain, 14–16 September, 2010; pp. 404–407.

15. Rosendale, G.; Kianian, S.; Manning, M.; Hamilton, D.; Huang, X.M.H.; Robinson, K.; Kim, Y.W.; Rueckes, T. A 4 Megabit Carbon Nanotube-based nonvolatile memory (NRAM). In Proceedings of the ESSDERC, Sevilla, Spain, 14–16 September, 2010; pp. 478–481.

16. Fuhrer, M.S.; Kim, B.M.; Durkop, T.; Brintlinger, T. High-mobility nanotube transistor memory. *Nano Lett.***2002**, *2*, 755–759.

17. Di Bartolomeo, A.; Yang, Y.; Rinzan, M.B.M.; Boyd, A.K.; Barbara, P. Record endurance for single-walled carbon nanotube-based memory cell. *Nanoscale Res. Lett.* **2010**, *5*, 1852–1855.

18. Jarosz, P.R.; Shaukat, A.; Schauerman, C.M.; Cress, C.D.; Kladitis, P.E.; Ridgley, R.D.; Landi, B.J. High-performance,

lightweight coaxial cable from carbon nanotube conductors. *ACS Appl. Mater. Interfaces* **2012**,*4*, 1103–1109.

19. Nair, P.R.; Alam, M.A. Performance limits of nanobiosensors. *Appl. Phys. Lett.* **2006**, *88*, 233120:1–233120:3.

20. Pengfei, Q.F.; Vermesh, O.; Grecu, M.; Javey, A.; Wang, O.; Dai, H.J.; Peng, S.; Cho, K.J. Toward large arrays of multiplex functionalized carbon nanotube sensors for highly sensitive and selective molecular detection. *Nano Lett.* **2003**, *3*, 347–351.

21. Collins, P.G.; Bradley, K.; Ishigami, M.; Zettl, A. Extreme oxygen sensitivity of electronic properties of carbon nanotubes. *Science* **2000**, *287*, 1801–1804.

22. Alam, M. Reliability- and process-variation aware design of integrated circuits. *Microelectron. Reliab.* **2008**,*48*, 1114–1122.

23. Lee, Y.H.; McMahon, W.; Mielke, N.; Lu, Y.L.R.; Walstra, S. Managing bias-temperature instability for product reliability. *Proc. Int. Symp. VLSI Technol.* **2007**, *1*, 52–53.

24. Moore, G.E. Cramming more components onto integrated circuits (Reprinted from Electronics, pg 114–117, April 19, 1965). *Proc. IEEE* **1998**, *86*, 82–85.

25. Kang, S.J.; Kocabas, C.; Ozel, T.; Shim, M.; Pimparkar, N.; Alam, M.A.; Rotkin, S.V.; Rogers, J.A. High-performance electronics using dense, perfectly aligned arrays of single-walled carbon nanotubes. *Nat. Nanotechnol.* **2007**, *2*, 230–236.

26. Patil, N.; Lin, A.; Myers, E.R.; Ryu, K.; Badmaev, A.; Zhou, C.W.; Wong, H.S.P.; Mitra, S. Wafer-scale growth and transfer of aligned single-walled carbon nanotubes. *IEEE Trans. Nanotechnol.* **2009**, *8*, 498–504.

27. Wang, C.; Badmaev, A.; Jooyaie, A.; Bao, M.Q.; Wang, K.L.; Galatsis, K.; Zhou, C.W. Radio frequency and linearity performance of transistors using high-purity semiconducting carbon nanotubes. *ACS Nano* **2011**,*5*, 4169–4176.

28. Cao, Q.; Han, S.-J.; Tulevski, G.S.; Franklin, A.D.; Haensch, W. Evaluation of field-effect mobility and contact resistance of transistors that use solution-processed single-walled carbon nanotubes. *ACS Nano***2012**, *6*, 6471–6477.

29. Vijayaraghavan, A.; Blatt, S.; Weissenberger, D.; Oron-Carl, M.; Hennrich, F.; Gerthsen, D.; Hahn, H.; Krupke, R. Ultra-large-scale directed assembly of single-walled carbon nanotube devices. *Nano Lett.* **2007**,*7*, 1556–1560.

30. Bardecker, J.A.; Afzali, A.; Tulevski, G.S.; Graham, T.; Hannon, J.B.; Jen, A.K.Y. Directed assembly of single-walled carbon nanotubes via drop-casting onto a UV-patterned photosensitive monolayer. *J. Am. Chem. Soc.* **2008**, *130*, 7226–7227.

31. Park, H.; Afzali, A.; Han, S.-J.; Tulevski, G.S.; Franklin, A.D.; Tersoff, J.; Hannon, J.B.; Haensch, W. High-density integration of carbon nanotubes via chemical self-assembly. *Nat. Nanotechnol.* **2012**, *7*, 787–791.

32. Cao, J.; Nyffeler, C.; Lister, K.; Ionescu, A.M. Resist-assisted assembly of single-walled carbon nanotube devices with nanoscale precision. *Carbon* **2012**, *50*, 1720–1726.

33. Wang, J. Carbon-nanotube based electrochemical biosensors: A review. *Electroanalysis* **2005**, *17*, 7–14.

34. Javey, A.; Guo, J.; Farmer, D.B.; Wang, Q.; Wang, D.W.; Gordon, R.G.; Lundstrom, M.; Dai, H.J. Carbon nanotube field-effect transistors with integrated ohmic contacts and high-k gate dielectrics. *Nano Lett.* **2004**,*4*, 447–450.

35. Cao, Q.; Han, S.-J.; Tulevski, G.S.; Zhu, Y.; Lu, D.D.; Haensch, W. Arrays of single-walled carbon nanotubes with full surface

coverage for high-performance electronics. *Nat. Nanotechnol.* **2013**, *8*, 180–186.

36. Franklin, A.D.; Wong, H.S.P.; Lin, A.; Chen, Z.H. Current scaling in aligned carbon nanotube array transistors with local bottom gating. *IEEE Electron Device Lett.* **2010**, *31*, 644–646.

37. Cao, J.; Ionescu, A.M. Self-aligned lateral dual-gate suspended-body single-walled carbon nanotube field-effect transistors. *Appl. Phys. Lett.* **2012**, *100*, 063103:1–063103:3.

38. Franklin, A.D.; Koswatta, S.O.; Farmer, D.B.; Smith, J.T.; Gignac, L.; Breslin, C.M.; Han, S.-J.; Tulevski, G.S.; Miyazoe, H.; Haensch, W.; *et al.* Carbon nanotube complementary wrap-gate transistors. *Nano Lett.* **2013**,*13*, 2490–2495.

39. Pimparkar, N.; Alam, M.A. A "bottom-up" redefinition for mobility and the effect of poor tube-tube contact on the performance of CNT nanonet thin-film transistors. *IEEE Electron Device Lett.* **2008**, *29*, 1037–1039.

40. Pimparkar, N.; Guo, J.; Alam, M.A. Performance assessment of subpercolating nanobundle network thin-film transistors by an analytical model. *IEEE Trans. Electron Devices* **2007**, *54*, 637–644.

41. Raychowdhury, A.; de, V.K.; Kurtin, J.; Borkar, S.Y.; Roy, K.; Keshavarzi, A. Variation tolerance in a multichannel carbon-nanotube transistor for high-speed digital circuits. *IEEE Trans. Electron Devices* **2009**,*56*, 383–392.

42. Nessim, G.D. Properties, synthesis, and growth mechanisms of carbon nanotubes with special focus on thermal chemical vapor deposition. *Nanoscale* **2010**, *2*, 1306–1323.

43. Cao, Q.; Rogers, J.A. Ultrathin films of single-walled carbon nanotubes for electronics and sensors: A review of fundamental and applied aspects. *Adv. Mater.* **2009**, *21*, 29–53.

44. Islam, A.E.; Du, F.; Ho, X.; Jin, S.H.; Dunham, S.; Rogers, J.A. Effect of variations in diameter and density on the statistics of aligned array carbon-nanotube field effect transistors. *J. Appl. Phys.* **2012**, *111*, 054511:1–054511:7.

45. Salamat, S.; Ho, X.N.; Rogers, J.A.; Alam, M.A. Intrinsic performance variability in aligned array CNFETs.*IEEE Trans. Nanotechnol.* **2011**, *10*, 439–444.

46. Sarker, B.K.; Shekhar, S.; Khondaker, S.I. Semiconducting enriched carbon nanotube aligned arrays of tunable density and their electrical transport properties. *ACS Nano* **2011**, *5*, 6297–6305.

47. Kocabas, C.; Kim, H.-S.; Banks, T.; Rogers, J.A.; Pesetski, A.A.; Baumgardner, J.E.; Krishnaswamy, S.V.; Zhang, H. Radio frequency analog electronics based on carbon nanotube transistors. *Proc. Natl. Acad. Sci. USA* **2008**, *105*, 1405–1409.

48. Hong, S.W.; Banks, T.; Rogers, J.A. Improved density in aligned arrays of single-walled carbon nanotubes by sequential chemical vapor deposition on quartz. *Adv. Mater.* **2010**, *22*, 1826–1830.

49. Xiao, J.L.; Dunham, S.; Liu, P.; Zhang, Y.W.; Kocabas, C.; Moh, L.; Huang, Y.G.; Hwang, K.C.; Lu, C.; Huang, W.; *et al.* Alignment controlled growth of single-walled carbon nanotubes on quartz substrates.*Nano Lett.* **2009**, *9*, 4311–4319.

50. Arnold, M.S.; Green, A.A.; Hulvat, J.F.; Stupp, S.I.; Hersam, M.C. Sorting carbon nanotubes by electronic structure using density differentiation. *Nat. Nanotechnol.* **2006**, *1*, 60–65.

51. Heller, D.A.; Mayrhofer, R.M.; Baik, S.; Grinkova, Y.V.; Usrey, M.L.; Strano, M.S. Concomitant length and diameter separation of single-walled carbon nanotubes. *J. Am. Chem. Soc.* **2004**, *126*, 14567–14573.

52. Zheng, M.; Jagota, A.; Strano, M.S.; Santos, A.P.; Barone, P.; Chou, S.G.; Diner, B.A.; Dresselhaus, M.S.; McLean, R.S.; Onoa, G.B.; *et al.* Structure-based carbon nanotube sorting by sequence-dependent DNA assembly. *Science* **2003**, *302*, 1545–1548.

53. Stokes, P.; Khondaker, S.I. High quality solution processed carbon nanotube transistors assembled by dielectrophoresis. *Appl. Phys. Lett.* **2010**, *96*, 083110:1–083110:3.

54. Jin, S.H.; Dunham, S.N.; Song, J.; Xie, X.; Kim, J.-H.; Lu, C.; Islam, A.; Du, F.; Kim, J.; Felts, J.; *et al.* Using nanoscale thermocapillary flows to create arrays of purely semiconducting single-walled carbon nanotubes. *Nat. Nanotechnol.* **2013**, *8*, 347–355.

55. Shulaker, M.M.; Wei, H.; Patil, N.; Provine, J.; Chen, H.Y.; Wong, H.S.P.; Mitra, S. Linear increases in carbon nanotube density through multiple transfer technique. *Nano Lett.* **2011**, *11*, 1881–1886.

56. Wang, C.; Ryu, K.; de Arco, L.G.; Badmaev, A.; Zhang, J.; Lin, X.; Che, Y.; Zhou, C. Synthesis and device applications of high-density aligned carbon nanotubes using low-pressure chemical vapor deposition and stacked multiple transfer. *Nano Res.* **2010**, *3*, 831–842.

57. Kim, W.; Javey, A.; Tu, R.; Cao, J.; Wang, Q.; Dai, H.J. Electrical contacts to carbon nanotubes down to 1 nm in diameter. *Appl. Phys. Lett.* **2005**, *87*, 173101:1–173101:3.

58. Tseng, Y.C.; Phoa, K.; Carlton, D.; Bokor, J. Effect of diameter variation in a large set of carbon nanotube transistors. *Nano Lett.* **2006**, *6*, 1364–1368.

59. Chen, Z.H.; Appenzeller, J.; Knoch, J.; Lin, Y.M.; Avouris, P. The role of metal-nanotube contact in the performance of carbon

nanotube field-effect transistors. *Nano Lett.* **2005**, *5*, 1497–1502.

60. Nosho, Y.; Ohno, Y.; Kishimoto, S.; Mizutani, T. n-Type carbon nanotube field-effect transistors fabricated by using Ca contact electrodes. *Appl. Phys. Lett.* **2005**, *86*, 073105:1–073105:3.

61. Wang, S.; Zeng, Q.S.; Yang, L.J.; Zhang, Z.Y.; Wang, Z.X.; Pei, T.A.; Ding, L.; Liang, X.L.; Gao, M.; Li, Y.; *et al.* High-performance carbon nanotube light-emitting diodes with Asymmetric contacts. *Nano Lett.* **2011**, *11*, 23–29.

62. Wei, H.; Chen, H.-Y.; Liyanage, L.; Wong, H.S.P.; Mitra, S. Air-Stable Technique for Fabricating n-Type Carbon Nanotube FETs. In Proceedings of the IEEE International Electron Devices Meeting, Washington DC, USA, 5–7 December, 2011; pp. 505–508.

63. Franklin, A.D.; Chen, Z.H. Length scaling of carbon nanotube transistors. *Nat. Nanotechnol.* **2010**, *5*, 858–862.

64. Franklin, A.D.; Luisier, M.; Han, S.-J.; Tulevski, G.; Breslin, C.M.; Gignac, L.; Lundstrom, M.S.; Haensch, W. Sub-10 nm carbon nanotube transistor. *Nano Lett.* **2012**, *12*, 758–762.

65. Lin, Y.M.; Appenzeller, J.; Avouris, P. Novel Carbon Nanotube FET Design with Tunable Polarity. In Proceedings of the IEEE International Electron Devices Meeting, SanFrancisco, CA, 13–15 December, 2004; pp. 687–690.

66. Sze, S.M. *Physics of Semiconductor Devices*, 2nd ed.; John Wiley & Sons: Hoboken, NY, USA, 1981.

67. Ho, X.; Ye, L.; Rotkin, S.V.; Cao, Q.; Unarunotai, S.; Salamat, S.; Alam, M.A.; Rogers, J.A. Scaling properties in transistors that use aligned arrays of single-walled carbon nanotubes. *Nano Lett.* **2010**, *10*, 499–503.

68. Deshpande, V.V.; Chandra, B.; Caldwell, R.; Novikov, D.S.; Hone, J.; Bockrath, M. Mott insulating state in ultraclean carbon nanotubes. *Science* **2009**, *323*, 106–110.

69. Charlier, J.C.; Blase, X.; Roche, S. Electronic and transport properties of nanotubes. *Rev. Mod. Phys.* **2007**, *79*, 677–732.

70. Wahab, M.A.; Jin, S.H.; Islam, A.E.; Kim, J.; Kim, J.-H.; Yeo, W.-H.; Lee, D.J.; Chung, H.U.; Rogers, J.A.; Alam, M.A. Electrostatic dimension of aligned-array carbon nanotube field-effect transistors. *ACS Nano***2013**, *7*, 1299–1308.

71. Rutherglen, C.; Jain, D.; Burke, P. Nanotube electronics for radiofrequency applications. *Nat. Nanotechnol.***2009**, *4*, 811–819.

72. Ryu, K.; Badmaev, A.; Wang, C.; Lin, A.; Patil, N.; Gomez, L.; Kumar, A.; Mitra, S.; Wong, H.S.P.; Zhou, C.W. CMOS-analogous wafer-scale nanotube-on-insulator approach for submicrometer devices and integrated circuits using aligned nanotubes. *Nano Lett.* **2009**, *9*, 189–197.

73. Ishikawa, F.N.; Chang, H.K.; Ryu, K.; Chen, P.C.; Badmaev, A.; de Arco, L.G.; Shen, G.Z.; Zhou, C.W. Transparent electronics based on transfer printed aligned carbon nanotubes on rigid and flexible substrates. *ACS Nano* **2009**, *3*, 73–79.

74. Feng, C.; Liu, K.; Wu, J.S.; Liu, L.; Cheng, J.S.; Zhang, Y.Y.; Sun, Y.H.; Li, Q.Q.; Fan, S.S.; Jiang, K.L. Flexible, stretchable, transparent conducting films made from superaligned carbon nanotubes. *Adv. Funct. Mater.***2010**, *20*, 885–891.

75. Ago, H.; Imamoto, K.; Ishigami, N.; Ohdo, R.; Ikeda, K.-I.; Tsuji, M. Competition and cooperation between lattice-oriented growth and step-templated growth of aligned carbon nanotubes on sapphire. *Appl. Phys. Lett.* **2007**, *90*, 123112:1–123112:3.

76. Ismach, A.; Segev, L.; Wachtel, E.; Joselevich, E. Atomic-step-templated formation of single wall carbon nanotube patterns. *Angew. Chem. Int. Ed.* **2004**, *43*, 6140–6143.

77. Kocabas, C.; Kang, S.J.; Ozel, T.; Shim, M.; Rogers, J.A. Improved synthesis of aligned arrays of single-walled carbon nanotubes and their implementation in thin film type transistors. *J. Phys. Chem. C* **2007**, *111*, 17879–17886.

78. Huang, H.J.; Maruyama, R.; Noda, K.; Kajiura, H.; Kadono, K. Preferential destruction of metallic single-walled carbon nanotubes by laser irradiation. *J. Phys. Chem. B* **2006**, *110*, 7316–7320.

79. Collins, P.C.; Arnold, M.S.; Avouris, P. Engineering carbon nanotubes and nanotube circuits using electrical breakdown. *Science* **2001**, *292*, 706–709.

80. Chen, C.X.; Yan, L.J.; Kong, E.S.W.; Zhang, Y.F. Ultrasonic nanowelding of carbon nanotubes to metal electrodes. *Nanotechnology* **2006**, *17*, 2192–2197.

81. Perebeinos, V.; Avouris, P. Impact excitation by hot carriers in carbon nanotubes. *Phy. Rev. B* **2006**, *74*.

82. Liao, A.; Zhao, Y.; Pop, E. Avalanche-induced current enhancement in semiconducting carbon nanotubes.*Phys. Rev. Lett.* **2008**, *101*, 256804.

83. Kim, W.; Javey, A.; Vermesh, O.; Wang, O.; Li, Y.M.; Dai, H.J. Hysteresis caused by water molecules in carbon nanotube field-effect transistors. *Nano Lett.* **2003**, *3*, 193–198.

84. Polishchuk, I.; Ranade, P.; King, T.J.; Hu, C.M. Dual work function metal gate CMOS technology using metal interdiffusion. *IEEE Electron Device Lett.* **2001**, *22*, 444–446.

85. Lim, A.E.J.; Hwang, W.S.; Wang, X.P.; Lai, D.M.Y.; Samudra, G.S.; Kwong, D.L.; Yeo, Y.C. Metal-gate work function modulation using hafnium alloys obtained by the

interdiffusion of thin metallic layers. *J. Electrochem. Soc.* **2007**, *154*, H309–H313.

86. Franklin, A.D.; Tulevski, G.S.; Han, S.-J.; Shahrjerdi, D.; Cao, Q.; Chen, H.-Y.; Wong, H.S.P.; Haensch, W. Variability in carbon nanotube transistors: Improving device-to-device consistency. *ACS Nano* **2012**, *6*, 1109–1115.

87. Paul, B.C.; Fujita, S.; Okajima, M.; Lee, T.H.; Wong, H.S.P.; Nishi, Y. Impact of a process variation on nanowire and nanotube device performance. *IEEE Trans. Electron Devices* **2007**, *54*, 2369–2376.

88. Wong, H.S.P.; Mitra, S.; Akinwande, D.; Beasley, C.; Chai, Y.; Chen, H.-Y.; Chen, X.; Close, G.; Deng, J.; Hazeghi, A.; *et al.* Carbon Nanotube Electronics—Materials, Devices, Circuits, Design, Modeling, and Performance Projection. In Proceedings of the IEEE International Electron Devices Meeting, Washington DC, USA, 5–7 December, 2011; pp. 501–504.

89. Balijepalli, A.; Sinha, S.; Cao, Y. Compact Modeling of Carbon Nanotube Transistor for Early Stage Process-Design Exploration. In Proceedings of the 2007 ACM/IEEE International Symposium on Low Power Electronics and Design (ISLPED), Portland, OR, USA, 27–29 August 2007.

90. Zhang, G.; Qi, P.; Wang, X.; Lu, Y.; Li, X.; Tu, R.; Bangsaruntip, S.; Mann, D.; Zhang, L.; Dai, H. Selective etching of metallic carbon nanotubes by gas-phase reaction. *Science* **2006**, *314*, 974–977.

91. Zhang, J.; Patil, N.P.; Hazeghi, A.; Wong, H.S.P.; Mitra, S. Characterization and design of Logic circuits in the presence of carbon nanotube density variations. *IEEE Trans. Comput. Aided Des. Integr. Circuits Syst.* **2011**, *30*, 1103–1113.

92. Boslaugh, S.; Watters, P.A. Safari Tech Books Online. In *Statistics in a Nutshell*; O'Reilly Media: Sebastopol, CA, USA, 2008.

93. Sun, Y.; Tuleski, G.; Han, S.-J.; Haensch, W.; Chen, Z. Improve Variability in Carbon Nanotube FETs by Scaling. In Proceedings of the Device Research Conference, NotreDame, IN, USA, 21–23 June, 2010; pp. 283–284.

94. Jin, S.H.; Islam, A.E.; Kim, T.-I.; Kim, J.-H.; Alam, M.A.; Rogers, J.A. Sources of hysteresis in carbon nanotube field-effect transistors and their elimination via methylsiloxane encapsulants and optimized growth. *Adv. Funct. Mater.* **2012**, *22*, 2276–2284.

95. Qu, M.; Qiu, Z.-J.; Zhang, Z.-B.; Li, H.; Li, J.; Zhang, S.-L. Charge-injection-induced time decay in carbon nanotube network-based FETs. *IEEE Electron Device Lett.* **2010**, *31*, 1098–1100.

96. Peng, N.; Zhang, Q.; Yuan, S.; Li, H.; Tian, J.; Chan, L. Current instability of carbon nanotube field effect transistors. *Nanotechnology* **2007**, *18*.

97. Ellison, M.D.; Crotty, M.J.; Koh, D.; Spray, R.L.; Tate, K.E. Adsorption of NH_3 and NO_2 on single-walled carbon nanotubes. *J. Phys. Chem. B* **2004**, *108*, 7938–7943.

98. Yao, Z.; Postma, H.W.C.; Balents, L.; Dekker, C. Carbon nanotube intramolecular junctions. *Nature* **1999**,*402*, 273–276.

99. Robinson, J.A.; Snow, E.S.; Badescu, S.C.; Reinecke, T.L.; Perkins, F.K. Role of defects in single-walled carbon nanotube chemical sensors. *Nano Lett.* **2006**, *6*, 1747–1751.

100. Zhou, C.W.; Kong, J.; Yenilmez, E.; Dai, H.J. Modulated chemical doping of individual carbon nanotubes.*Science* **2000**, *290*, 1552–1555.

101. Zhuravlev, L.T. The surface chemistry of amorphous silica. Zhuravlev model. *Colloids Surfaces Physicochem. Eng. Asp.* **2000**, *173*, 1–38.

102. Chakrapani, V.; Anderson, A.B.; Wolter, S.D.; Stoner, B.R.; Sumanasekera, G.U.; Angus, J.C. Charge transfer equilibria between diamond and an aqueous oxygen electrochemical redox couple. *Science* **2007**,*318*, 1424–1430.

103. McGill, S.A.; Rao, S.G.; Manandhar, P.; Xiong, P.; Hong, S. High-performance, hysteresis-free carbon nanotube field-effect transistors via directed assembly. *Appl. Phys. Lett.* **2006**, *89*, 163123:1–163123:3.

104. Aguirre, C.M.; Levesque, P.L.; Paillet, M.; Lapointe, F.; St-Antoine, B.C.; Desjardins, P.; Martel, R. The role of the oxygen/water redox couple in suppressing electron conduction in field-effect transistors. *Adv. Mater.***2009**, *21*.

105. Levesque, P.L.; Sabri, S.S.; Aguirre, C.M.; Guillemette, J.; Siaj, M.; Desjardins, P.; Szkopek, T.; Martel, R. Probing charge transfer at surfaces using graphene transistors. *Nano Lett.* **2011**, *11*, 132–137.

106. Weitz, R.T.; Zschieschang, U.; Effenberger, F.; Klauk, H.; Burghard, M.; Kern, K. High-performance carbon nanotube field effect transistors with a thin gate dielectric based on a self-assembled monolayer. *Nano Lett.***2007**, *7*, 22–27.

107. Hur, S.H.; Yoon, M.H.; Gaur, A.; Shim, M.; Facchetti, A.; Marks, T.J.; Rogers, J.A. Organic nanodielectrics for low voltage carbon nanotube thin film transistors and complementary logic gates. *J. Am. Chem. Soc.***2005**, *127*, 13808–13809.

108. Reisman, A.; Williams, C.K.; Maldonado, J.R. Generation and annealing of defects in silicon dioxide. *J. Appl. Phys.* **1987**, *62*, 868–874.

109. Alam, M.A. SILC as a measure of trap generation and predictor of T-BD in ultrathin oxides. *IEEE Trans. Electron Devices* **2002**, *49*, 226–231.

110. Solomon, P. Breakdown in silicon-oxide. *J. Vac. Sci. Technol.* **1977**, *14*, 1122–1130.

111. Ribes, G.; Mitard, J.; Denais, M.; Bruyere, S.; Monsieur, F.; Parthasarathy, C.R.; Vincent, E.; Ghibaudo, G. Review on high-k dielectrics reliability issues. *IEEE Trans. Device Mater Reliab.* **2005**, *5*, 5–19.

112. Lenahan, P.M.; Dressendorfer, P.V. Hole traps and trivalent silicon centers in metal-oxide silicon devices. *J. Appl. Phys.* **1984**, *55*, 3495–3499.

113. Lenahan, P.M. Dominating defects in the MOS system: P_b and E' centers. In *Defects in Microelectronic Materials and Devices*; Fleetwood, S.P.D., Schrimpf, R.D., Eds.; CRC Press: Taylor and Francis Group, Boca Raton, FL, USA, 2009; pp. 163–214.

114. Skuja, L. Optically active oxygen-deficiency-related centers in amorphous silicon dioxide. *J. Non Cryst. Solids* **1998**, *239*, 16–48.

115. Spinelli, A.S.; Lacaita, A.L.; Rigamonti, M.; Ielmini, D.; Ghidini, G. Separation of electron and hole traps by transient current analysis. *Microelectron. Eng.* **1999**, *48*, 151–154.

116. Hongo, H.; Nihey, F.; Yorozu, S. Relationship between carbon nanotube density and hysteresis characteristics of carbon nanotube random network-channel field effect transistors. *J. Appl. Phys.* **2010**, *107*, 094501:1–094501:9.

117. Di Bartolomeo, A.; Rinzan, M.; Boyd, A.K.; Yang, Y.; Guadagno, L.; Giubileo, F.; Barbara, P. Electrical properties and memory effects of field-effect transistors from networks of single- and double-walled carbon nanotubes. *Nanotechnology* **2010**, *21*.

118. Memming, R. *Semiconductor Electrochemistry*; Wiley-VCH: Weinheim, NY, USA, 2001.

119. Shockley, W.; Read, W.T. Statistics of the recombinations of holes and electrons. *Phys. Rev.* **1952**, *87*, 835–842.

120. Chang, M.H.; Zhang, J.F.; Zhang, W.D. Assessment of capture cross sections and effective density of electron traps generated in silicon dioxides. *IEEE Trans. Electron Devices* **2006**, *53*, 1347–1354.

121. Hu, P.; Zhang, C.; Fasoli, A.; Scardaci, V.; Pisana, S.; Hasan, T.; Robertson, J.; Milne, W.I.; Ferrari, A.C. Hysteresis suppression in self-assembled single-wall nanotube field effect transistors. *Phys. ELowDimens. Syst. Nanostruct.* **2008**, *40*, 2278–2282.

122. Kim, S.K.; Xuan, Y.; Ye, P.D.; Mohammadi, S.; Back, J.H.; Shim, M. Atomic layer deposited Al_2O_3 for gate dielectric and passivation layer of single-walled carbon nanotube transistors. *Appl. Phys. Lett.* **2007**, *90*, 163108:1–163108:3.

123. Kirton, M.J.; Uren, M.J. Noise in solid-state microstructures—A new perspective on individual defects, interface states and low-frequency (1/f) noise. *Adv. Phys.* **1989**, *38*, 367–468.

124. Vandamme, L.K.J. Noise as a diagnostic tool for quality and reliability of electronic devices. *IEEE Trans. Electron Devices* **1994**, *41*, 2176–2187.

125. Vandamme, L.K.J.; Li, X.S.; Rigaud, D. 1/f Noise in MOS Devices, Mobility or Number Fluctuations. *IEEE Trans. Electron Devices* **1994**, *41*, 1936–1945.

126. Weissman, M.B. 1/f noise and other slow, nonexponential kinetics in condensed matter. *Rev. Mod. Phys.***1988**, *60*, 537–571.

127. Weissman, M.B. Low-frequency noise as a tool to study disordered materials. *Annu. Rev. Mater. Sci.* **1996**,*26*, 395–429.

128. Kurata, H.; Otsuga, K.; Kotabe, A.; Kajiyama, S.; Osabe, T.; Sasago, Y.; Narumi, S.; Tokami, K.; Kamohara, S.; Tsuchiya, O. Random telegraph signal in flash memory: Its impact on scaling of multilevel flash memory beyond the 90-nm node. *IEEE J. SolidState Circuits* **2007**, *42*, 1362–1369.

129. Ghetti, A.; Compagnoni, C.M.; Biancardi, F.; Lacaita, A.L.; Beltrami, S.; Chiavarone, L.; Spinelli, A.S.; Visconti, A. Scaling trends for random telegraph noise in deca-nanometer Flash memories. In Proceedings of the IEEE International Electron Devices Meeting Technical Digest, SanFrancisco, CA, 15–17 December, 2008; pp. 835–838.

130. Gierkink, S.L.J.; Klumperink, E.A.M.; van der Wel, A.P.; Hoogzaad, G.; van Tuijl, E.; Nauta, B. Intrinsic 1/f device noise reduction and its effect on phase noise in CMOS ring oscillators. *IEEE J. SolidState Circuits***1999**, *34*, 1022–1025.

131. Cao, Q.; Xia, M.G.; Kocabas, C.; Shim, M.; Rotkin, S.V.; Rogers, J.A. Gate capacitance coupling of singled-walled carbon nanotube thin-film transistors. *Appl. Phys. Lett.* **2007**, *90*, 023516:1–023516:3.

132. Rosenblatt, S.; Yaish, Y.; Park, J.; Gore, J.; Sazonova, V.; McEuen, P.L. High performance electrolyte gated carbon nanotube transistors. *Nano Lett.* **2002**, *2*, 869–872.

133. Vitusevich, S.A.; Sydoruk, V.A.; Petrychuk, M.V.; Danilchenko, B.A.; Klein, N.; Offenhaeusser, A.; Ural, A.; Bosman, G. Transport properties of single-walled carbon nanotube transistors after gamma radiation treatment. *J. Appl. Phys.* **2010**, *107*, 063701:1–063701:6.

134. Lin, Y.-M.; Appenzeller, J.; Knoch, J.; Chen, Z.; Avouris, P. Low-frequency current fluctuations in individual semiconducting single-wall carbon nanotubes. *Nano Lett.* **2006**, *6*, 930–936.

135. Collins, P.G.; Fuhrer, M.S.; Zettl, A. 1/f noise in carbon nanotubes. *Appl. Phys. Lett.* **2000**, *76*, 894–896.

136. Snow, E.S.; Novak, J.P.; Lay, M.D.; Perkins, F.K. 1/f noise in single-walled carbon nanotube devices. *Appl. Phys. Lett.* **2004**, *85*, 4172–4174.

137. Liu, F.; Wang, K.L.; Li, C.; Zhou, C. Study of random telegraph signals in single-walled carbon nanotube field effect transistors. *IEEE Trans. Nanotechnol.* **2006**, *5*, 441–445.

138. Vitusevich, S.; Gasparyan, F. Low-Frequency Noise Spectroscopy at Nanoscale: Carbon Nanotube Materials and Devices. Available online: http://cdn.intechweb.org/pdfs/17294.pdf (accessed on 1 July 2013).

139. Veksler, D.; Bersuker, G.; Rumyantsev, S.; Shur, M.; Park, H.; Young, C.; Lim, K.Y.; Taylor, W.; Jammy, R.; IEEE. Understanding noise measurements in MOSFETs: The role of traps structural relaxation. In Proceedings of the International Reliability Physics Symposium, Anaheim, CA, 2–6 May, 2010; pp. 73–79.

140. Hooge, F.N. 1/F noise sources. *IEEE Trans. Electron Devices* **1994**, *41*, 1926–1935.

141. Liu, F.; Bao, M.; Wang, K.L.; Zhang, D.; Zhou, C. Coulomb attractive random telegraph signal in a single-walled carbon nanotube. *Phys. Rev. B* **2006**, *74*.

142. Liu, F.; Wang, K.L. Correlated random telegraph signal and low-frequency noise in carbon nanotube transistors. *Nano Lett.* **2008**, *8*, 147–151.

143. Liu, F.; Wang, K.L.; Zhang, D.; Zhou, C. Random telegraph signals and noise behaviors in carbon nanotube transistors. *Appl. Phys. Lett.* **2006**, *89*, 243101:1–243101:3.

144. Chan, J.; Kidd, D.; Burke, B.; Harriott, L.; Williams, K. Noise analysis of carbon nanotube field effect transistors irradiated by electron beam. *J. Vac. Sci. Technol. B* **2010**, *28*, C6P66–C6P69.

145. Cress, C.D.; McMorrow, J.J.; Robinson, J.T.; Friedman, A.L.; Hughes, H.L.; Weaver, B.D.; Landi, B.J. Total ionizing dose-hardened carbon nanotube thin-film transistors with silicon oxynitride gate dielectrics. *MRS Commun.* **2011**, *1*, 27–31.

146. Cress, C.D.; McMorrow, J.J.; Robinson, J.T.; Friedman, A.L.; Landi, B.J. Radiation effects in single-walled carbon nanotube thin-film-transistors. *IEEE Trans. Nucl. Sci.* **2010**, *57*, 3040–3045.

147. Hulman, M.; Skakalova, V.; Roth, S.; Kuzmany, H. Raman spectroscopy of single-wall carbon nanotubes and graphite irradiated by gamma rays. *J. Appl. Phys.* **2005**, *98*, 024311:1–024311:5.

148. Rossi, J.E.; Cress, C.D.; Helenic, A.R.; Schauerman, C.M.; DiLeo, R.A.; Cox, N.D.; Messenger, S.R.; Weaver, B.D.; Hubbard, S.M.; Landi, B.J. Ion irradiation of electronic-type-separated single wall carbon nanotubes: A model for radiation effects in nanostructured carbon. *J. Appl. Phys.* **2012**, *112*, 034314:1–034314:11.

149. Tang, X.W.; Yang, Y.; Kim, W.; Wang, Q.; Qi, P.F.; Dai, H.J.; Xing, L. Measurement of ionizing radiation using carbon nanotube field effect transistor. *Phys. Med. Biol.* **2005**, *50*, N23–N31.

150. Cress, C.D.; Schauerman, C.M.; Landi, B.J.; Messenger, S.R.; Raffaelle, R.P.; Walters, R.J. Radiation effects in single-walled

carbon nanotube papers. *J. Appl. Phys.* **2010**, *107*, 014316:1–014316:5.

151. Skakalova, V.; Woo, Y.S.; Osvath, Z.; Biro, L.P.; Roth, S. Electron transport in Ar+-irradiated single wall carbon nanotubes. *Phys. Status Solidi B* **2006**, *243*, 3346–3350.

152. Dresselhaus, M.S.; Dresselhaus, G.; Saito, R.; Jorio, A. Raman spectroscopy of carbon nanotubes. *Phys. Rep. Rev. Sect. Phys. Lett.* **2005**, *409*, 47–99.

153. Buchowicz, G.; Stone, P.R.; Robinson, J.T.; Cress, C.D.; Beeman, J.W.; Dubon, O.D. Correlation between structure and electrical transport in ion-irradiated graphene grown on Cu foils. *Appl. Phys. Lett.* **2010**, *98*, 032102:1–032102:3.

154. Hong, W.-K.; Lee, C.; Nepal, D.; Geckeler, K.E.; Shin, K.; Lee, T. Radiation hardness of the electrical properties of carbon nanotube network field effect transistors under high-energy proton irradiation.*Nanotechnology* **2006**, *17*, 5675–5680.

155. Hughes, H.L.; Benedetto, J.M. Radiation effects and hardening of MOS technology: Devices and circuits.*IEEE Trans. Nucl. Sci.* **2003**, *50*, 500–521.

156. Islam, A.E.; Kufluoglu, H.; Varghese, D.; Mahapatra, S.; Alam, M.A. Recent issues in negative bias temperature instability: Initial degradation, field-dependence of interface trap generation, hole trapping effects, and relaxation. *IEEE Trans. Electron Devices* **2007**, *54*, 2143–2154.

157. Schroder, D.K.; Babcock, J.A. Negative bias temperature instability: Road to cross in deep submicron silicon semiconductor manufacturing. *J. Appl. Phys.* **2003**, *94*, 1–18.

158. Gu, G.; Kane, M.G.; Mau, S.C. Reversible memory effects and acceptor states in pentacene-based organic thin-film transistors. *J. Appl. Phys.* **2007**, *101*, 014504:1–014504:9.

159. Liu, J.; Wang, C.; Tu, X.; Liu, B.; Chen, L.; Zheng, M.; Zhou, C. Chirality-controlled synthesis of single-wall carbon nanotubes using vapour-phase epitaxy. *Nat. Commun.* **2012**, *3*.

160. Wilk, G.D.; Wallace, R.M.; Anthony, J.M. High-kappa gate dielectrics: Current status and materials properties considerations. *J. Appl. Phys.* **2001**, *89*, 5243–5275.

Chapter 6

Optical and Mechanical Study of Mineral and Synthetic Nano Layered Silicate Reinforced Polyurethane Resin

Behdad Ahmadi[1], Ehsan Rezazadeh Mafi[1], Tahereh Samaee Yekta[1], Iman Rezazadeh Mafi[2], Seyed Mahmoud Kasiriha[1]

[1]Polymer Engineering Department, Amirkabir University of Technology, Tehran, Iran

[2]Department of Metallurgy and Materials Engineering, Tehran University, Tehran, Iran

ABSTRACT

In the present investigation, the effect of two different mineral nano clays and a synthetic nano layered silicate on the properties of a transparent polyurethane resin has been studied. Both high rotator Torusmill and high intensity ultrasonic deagglomerator are utilized for the dispersion process. Observations by means of the X-ray diffraction technique indicate the presence of different dispersion status of nano layers in polymer matrix. The optical effect of nano-filled clear coat was studied using goniospectrophotometer and compared with a blank clear coat to determine if it can be identified by human eye. Scratch and mar resistance tests presented considerable improvement. Gloss retention against rubbing increased significantly in the presence of merely 3 wt% of synthetic nano-filler.

Keywords: Polymer-Matrix Composites (PMCs); Nano Composite; Scratch Resistance; Hardness; Optical Properties

1. INTRODUCTION

Due to the ongoing customer demand for higher quality and more durable coatings, there has been a surge of interest in improving the physical damage resistance of automotive refinish coatings [1]. Automotive clear coats are designed both to protect the substrate from corrosion and to provide glamour. Nevertheless, the topcoats of most multi-layer coating systems have very high gloss characteristics, making scratch damage highly visible [2]. Consequently, in addition to bulk properties, surface properties such as resistance to indentation and scratch, are critical to retain the basic coating functions in automotive coatings [3].

Several chemical approaches can be utilized to achieve scratch resistance of coatings [4]. Among these possibilities, are two main methods to improve the scratch resistance of organic coatings: one is to optimize the polymer lacquer components and the other is to reinforce the coatings by embedding fillers into them.

Nanopowders with particle sizes under 100 nm represent a promising group of fillers. Nanoparticles have been shown to improve the mechanical properties even at low loadings due to their minute size while not affecting the transparency of clear coats [5]. Fundamentally, Inorganic-organic composite materials are immensely important due to their extraordinary properties, which arise from the synergism between the components. These materials have gained much interest due to their remarkable physical and chemical properties, including thermal stability [6-8], mechanical properties [6,7,9-12], flame retardant [13-16], barrier resistance [6,7,17-19], solvent resistance [20] and electrorheology properties [21,22]. Recently, the application of polymer-clay nanocomposites is found to be used as advanced anticorrosion coatings, which has been reported by Yeh et al. [23-38].

A very common 2 dimensional nano filler is layered silicate. Phyllosilicates are small flakes. Each layer consists of two sheets of tetrahedral silica (corner shared) with an edge shared with octahedral sheets of either alumina (aluminosilicate) or magnesia (magnesium silicate). Due to isomorphic substitution of alumina into the layered silicates (Al^{3+} for Si^{4-}) or magnesium for aluminum (Mg^{2+} for Al^{3+}) each layer has a negative charge between 0.5 and 1.3. The layers are held together with a layer of charge-compensating cations such as Li^+ and Na^+. These charge-compensating cations provide a route to the rich intercalation chemistry and surface modification required to disperse layered silicates at the nano scale into polymer [39]. Layered silicates dispersed as a reinforcing phase in an engineering polymer matrix are emerging as a relatively new form of useful materials.

These composites exhibit a change in composition and structure over a nanometer length scale. Owing to the nanometer size particles obtained by dispersion, these nanocomposites exhibit superior properties such as mechanical, thermal, optical and physico-chemical properties at a lower level loading compared with either the pure polymer or conventional micron sized composites. Their unique properties stem from a combination of factors; namely the platelet structure of nanolayer layered silicate, the high aspect ratio (width to thickness) of the platelets with thicknesses of the order of a nanometer, widths and lengths of the order of 25 - 500 nm, and the molecular bonds formed between the platelets and the polymer during compounding that may modify polymer properties [40].

There are several procedures to incorporate inorganic layered 2D materials finely dispersed into a polymer matrix. In principal these can be distinguished using the following classification: 1) intercalated hybrids in which one or more molecular layers of polymer chains are inserted between the sheets or galleries of the inorganic host, 2) delaminated hybrids in which the layers are exfoliated (or delaminated) and homogeneously dispersed in a polymer matrix [41,42]. For these layered silicates to be useful as nano composite, the layers must be separated. Layered silicates are inherently hydrophilic, but polymers tend to be hydrophobic. Usually, the space between the layers is intercalated by alkylammonium ions. These cations lower the surface energy of the inorganic host and improve the wetting characteristic of the materials with regard to the polymer. This opens the galleries sufficiently and makes them hydrophobic enough for a polymer chain to intercalate between the layers [43].

The main objective of this study is to investigate the effect of well dispersed nano layered silicates on mechanical and optical properties of a polyurethane base resin aimed to be used for automotive refinish clear coat and optimizing nano-filler/polymer ratio considering other important characteristics of a clear coat like impact resistance, adhesion, gloss, etc.

2. EXPERIMENTAL

2.1. Materials

Cloisite® Na^+ (Southern Clay Products, Inc.) is used as an unmodified mineral nano clay with a moisture content of 6% and a 7% weight loss on ignition. The typical aggregated dry particle size of Cloisite® Na^+ is 10% less than 2 μ, 50% less than 6 μ and 90% less than 13 μ. Loose bulk density is 12.45 lbs/ft^3 while packed bulk and pure density are stated as 20.95 lbs/ft and 2.86 g/cc respectively. "d-value" parameter corresponds to 11.7 Å.

The Second mineral clay used in this study is Cloisite®30B. A natural montmorillonite modified with a quaternary ammonium salt (methyl, tallow, bis-2-hydroxyethyl while "T" as tallow consists of ≈65% C18; ≈30% C16; ≈5% C14). The modifier concentration is 90 meq/100g clay. This clay's moisture content is less than 2% and its weight loss on ignition is 30%. Typical dry particle size of Cloisite®30B is as follows: 10% less than 2 μ, 50% less than 6 μ and 90% less than 13 μ. Loose bulk density is 14.25 lbs/ft^3 where packed bulk and pure density are 22.71 lbs/ft and 1.98 g/cc. "d-value" parameter is stated as 18.5 Å.

The other nano filler used in this study is an organically modified synthetic nano layered silicate (Nanothix) with 1 nm × 50 nm dimensions at complete exfoliation and a maximum moisture content of approximately 5% (Sud Chemie Inc.). This nano layered silicate is modified with magnesium silicate.

The polyurethane used here is a 2K polyurethane automotive refinish clear coat based on a hydroxyl functional acrylic resin (65% solid content), manufactured by Khosh Paint & Resin, Inc. T.D.I (toluene diisocyanate) is used as hardener.

2.2. Sample Preparation

Thick mild steel panels of 0.8mm were used as substrate all of which were painted with a typical solid automotive white paint. Prior to film application, all the substrates were cleaned according to Sa-3 degree in DIN 55928 standard. Once spraying had been completed, paint films were dried for a duration of 2 hours at 60°C. The dry film thickness of white paint was 30 - 45 µm that was applied in whole from the very same batch.

After a 2-hr dehumidifying process of the synthetic layered silicate powder at 80°C in an oven, three sets of samples containing 1, 3 and 5 wt% of each nano filler were prepared as follows:

Initially 5, 15 and 25 grams of each nano layered silicate were added to 50 cc of Xylene respectively and mixed by a common high rotator mixer for 10 minutes at 1500 rpm and then further mixed using an ultrasonic deagglomerator (Hielscher Ultrasonics GmbH, Model: PU400S, 4200 watts, 24 kHz) for another 10 minutes while the container was placed in a cold bath. Then, 434 grams of 65% solid resin were added to the container and mixed by a Torusmill (Dispermat instruments) at 3500 rpm for 1.5 hrs. It is worth noting that although the case might be different according to instrument or accessories used, it is strongly recommended not to process the mixture more than 2 hrs using Torusmill due to its improper optical effect.

At the next step, the obtained mixture was subjected to ultrasonic waves (Sonotrode used: H14 made of titanium, tip diameter 14mm, approx. length 100 mm, male thread M10 × 1) for 30 minutes to reach its complementary deagglomeration. To avoid polymer degradation, a "3 second off-3 second on" run mode was chosen. Once the samples had been mixed sufficiently, a stoichiometric amount of T.D.I (weight ratio of 1:2 equal to 217 grams) was added to each sample and mixed with a spatula.

The clear coat samples were applied on pre-prepared white panels with a film applicator to ensure thickness unity. The dry film thickness of clear coat was about 30 µm. This way samples containing 1, 3 and 5 wt% of filler were prepared. The rheological behavior of the samples did not show any noticeable difference. In addition, a 30µm thickness clear coat was applied over a bare mild steel panel and also on a glass substrate using film applicator. For assessing the optical properties, 2 sets of samples (blank and filler contained clear coat) were sprayed over white paint coated panels. The dry film thicknesses of sprayed samples were 30 - 45 µm. Finally, the very same procedure was implemented for Cloisite® Na^+ and Cloisite®30B.

Bulk samples prepared using mineral clay had a slight khaki-yellowish shade while the synthetic layered silicate-contained mixture had a whitish shade.

2.3. Test Methods and Experimental

X-ray diffraction (XRD) was used to study the state of dispersion.

XRD is the most commonly used technique to identify intercalated structures due to the periodic arrangement of the silicate layers both in the pristine and the intercalated states. Intercalation of polymer chains increases the interlayer spacing in comparison to the spacing of the original silicate, resulting in a shift of the X-ray diffraction (XRD) peak towards lower angles. The diffraction angle is related to the layer spacing "d" through the well-known Bragg's equation $\lambda = 2d\sin\theta$, where λ is X ray wavelength and 2θ is the scattering peak position. Exfoliated structure leads to the disappearance of any coherent X-ray scattering from the layers [40].

Two different methods were adopted to monitor the scratch resistance of coatings. The first one, the Chipping test (Pars Horm Instrument) was performed according to ASTM D3170 standard on sprayed substrate. The test consists of projecting a ball coated with silicon carbide onto the surface of a coating applied to a support in order to reproduce the superficial degradation (scaling) equivalent to that caused by grit projected on a vehicle. The second one, the Automatic Scratch test (Sheen Instrument) was performed according to ASTM D5178. The pendulum hardness test (Erichsen Instrument) was performed according to ASTM D4366-95. This test method is intended to measure the hardness using a Persoz pendulum for coatings applied to plane substrates. The test is aimed at determining the damping of the oscillations of a pendulum resting on the film to be examined by two steel balls, of equal diameter, being integral with the pendulum (performed as from amplitude 12° down to amplitude 4°). The test was performed on samples after 48 hrs and also 1 week after application. The coating films were prepared on glass boards and dried at room temperature. Some complementary tests were performed to find out whether the addition of nano filler can adversely affect other important properties of coating or not. The adhesion of coating was measured by the pull-off method according to ASTM D4541 and ISO 4624 (DeFelsko Instrument). Pull-off method is a quantitative test for adhesion where a loading fixture, commonly called a dolly or stub, is affixed by an adhesive to a coating. The force required to pull the dolly off or the force the dolly withstamds, yields the tensile strength in pounds per square inch (psi) or mega Pascals (MPa). Failure will occur along the weakest plane within the system (comprised of the dolly, adhesive, coating system and substrate) and will be exposed by the fracture surface.

The gloss of the coatings was measured by Sheen Instrument according to ASTM D523. The loss in gloss after 70 double rubs of abrasive was also measured according to DIN 53778. Mandrel bending (Elcometer Instrument) and Impact (Elcometer Instrument) tests were also performed according to ASTM D4338 and ASTM D2794 respectively.

The effect of nano-filled clear coat on reflectance spectrum of a typical white solid paint was compared with the effect of non-filled polyurethane by Goniospectrophotometer (Geretag machbeth, 741 gl instrument) in 4 different angles under 3 different common light sources. The colorimeter is used in the differential mode and it directly indicates the colour variations in the L, a, b system (I.E.C. 1976). There are several illuminants widely used by the automotive industry which include A, D65 and TL84 [44]. Illuminant A represents tungsten light. Illuminant D65 closely resembles the relative spectral energy distribution of north-sky daylight. Illuminant TL84 is a custom fluorescent. The results were reported as ΔE where ΔE is the overall colorimetric deviation and is calculated as

$$\Delta E = \sqrt{\left(L_\alpha - L_\beta\right)^2 + \left(a_\alpha - a_\beta\right)^2 + \left(b_\alpha - b_\beta\right)^2}$$

where, L, a and b are color components.

CIE Lab allows the specification of color perceptions in terms of a three-dimensional space. The colour variation is expressed by the average values of the colorimetric standard deviations computed with respect to the specimen: Delta L (ΔL), Delta a (Δa) and Delta b (Δb) and the standard deviations obtained on these values if these are above 0.1. The L* axis is known as the lightness and extends from 0 (black) to 100 (white). The other two coordinates a* and b* represent redness-greenness and yellowness-blueness respectively. Samples for which a* = b* = 0 are achromatic and thus the L* axis represents the achromatic scale of grays from black to white. A comparison list regarding color components is presented below:

ΔL < 0: darker

Δa > 0: redder (or less green)

Δa < 0: greener (or less red)

Δb > 0: yellower (or less blue)

Δb < 0: bluer (or less yellow)

The quantities L*, a*, and b* are obtained from the tristimulus values according to the following transformations:

L* = 116(Y/Yn)1/3 – 16 a* = 500[(X/Xn)1/3 – (Y/Yn)1/3]

b* = 200[(Y/Yn)1/3 – (Z/Zn)1/3]

where Xn, Yn, and Zn are the values of X, Y, and Z for the illuminant that was used for the calculation of X, Y and Z of samples.

To measure the drying time, clear coat applied samples were cured at room temperature. The drying time recorder (Braive instrument) was employed according to ASTM D5895.

3. RESULTS AND DISCUSSION

In order to investigate the dispersion state of nano particles in resin matrix, X-ray diffraction was performed on nano particles powder and on the polymer free film containing dispersed particles. XRD test was implemented under the following conditions:

Anode material: Cu Start angle [°2Θ]: 2.010 End angle [°2Θ]: 49.990 Start d-value [Å]: 43.9176 End d-value [Å]: 1.8230

α1 wavelength [Å]: 1.54060

α2 wavelength [Å]: 1.54439 The XRD curve for Cloisite® Na^+ in powder form and after dispersion in polymer free film is shown in **Figure 1**.

The 2θ values along the horizontal axis can be converted to layer spacing value "d" through Bragg's equation (λ = 2d sinθ).

d-values for Cloisite® Na^+ powder are as follows:

2Θ = 35.5 corresponding to 2.5 Å interlayer distance 2Θ = 28 corresponding to 3.1 Å interlayer distance 2Θ = 19 corresponding to 4.4 Å interlayer distance 2Θ = 7 corresponding to 11.4 Å interlayer distance As clarified above, four main peaks can be recognized through the XRD curve of Cloisite® Na^+ powder. However, taking into account the intensity of peaks (Y-axis), 2Θ = 28, 2Θ = 19 and 2Θ = 7 are main peaks having almost same intensity.

The "d-values" for processed sample obtained from free film are as follows:

2Θ = 19 corresponding to 4.4 Å interlayer distance 2Θ = 10 corresponding to 8.6 Å interlayer distance 2Θ = 5 corresponding to 16.5 Å interlayer distance As it is clear in **Figure 1**(b), after processing the nano filler, the three main peaks that occurred in Cloisite® Na^+ powder merged into a peak at 2Θ = 19. In addition, a new peak at 2Θ = 5 has appeared. However, the intensity of the peak is not quite noticeable.

Overall, the results show a clear increase in the interlayer distance of Cloisite® Na+ after process although it is not sufficient to be considered an exfoliated nano composite. In fact, the Cloisite® Na^+ final status can be interpreted as a partially intercalated nano composite.

Figure 2 shows the XRD results of Cloisite®30B; the organically modified nano clay used in this study.

Analyzing the main peaks of Cloisite®30B presents the following:

d-values for Cloisite®30B powder:

2Θ = 35.5 corresponding to 2.5 Å interlayer distance 2Θ = 28 corresponding to 3.1 Å interlayer distance 2Θ = 19 corresponding to 4.4 Å interlayer distance 2Θ = 10 corresponding to 8.6 Å interlayer distance 2Θ = 5 corresponding to 16.5 Å interlayer distance The position of peaks in case of Cloisite®30B resembles Cloisite® Na^+. However, the intensity of the last peak (2Θ = 5) is quite higher than other peaks and thus can be regarded as the main dominant peak of this sample.

d-values for processed Cloisite® 30B sample obtained from free film:

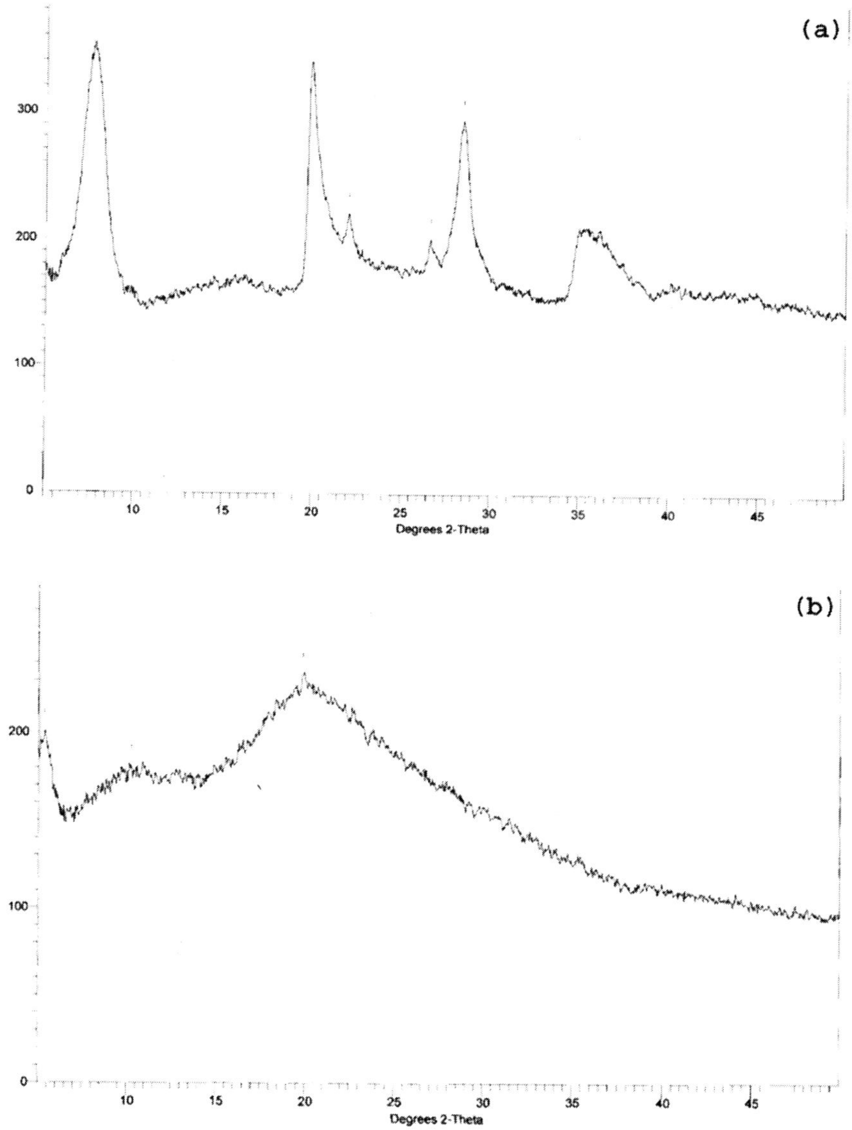

Figure 1. X-ray diffraction analysis showing the diffraction patterns that result from (a) Cloisite® Na^+ in powder form, (b) 3 wt% of Cloisite® Na^+ after dispersion in polyurethane.

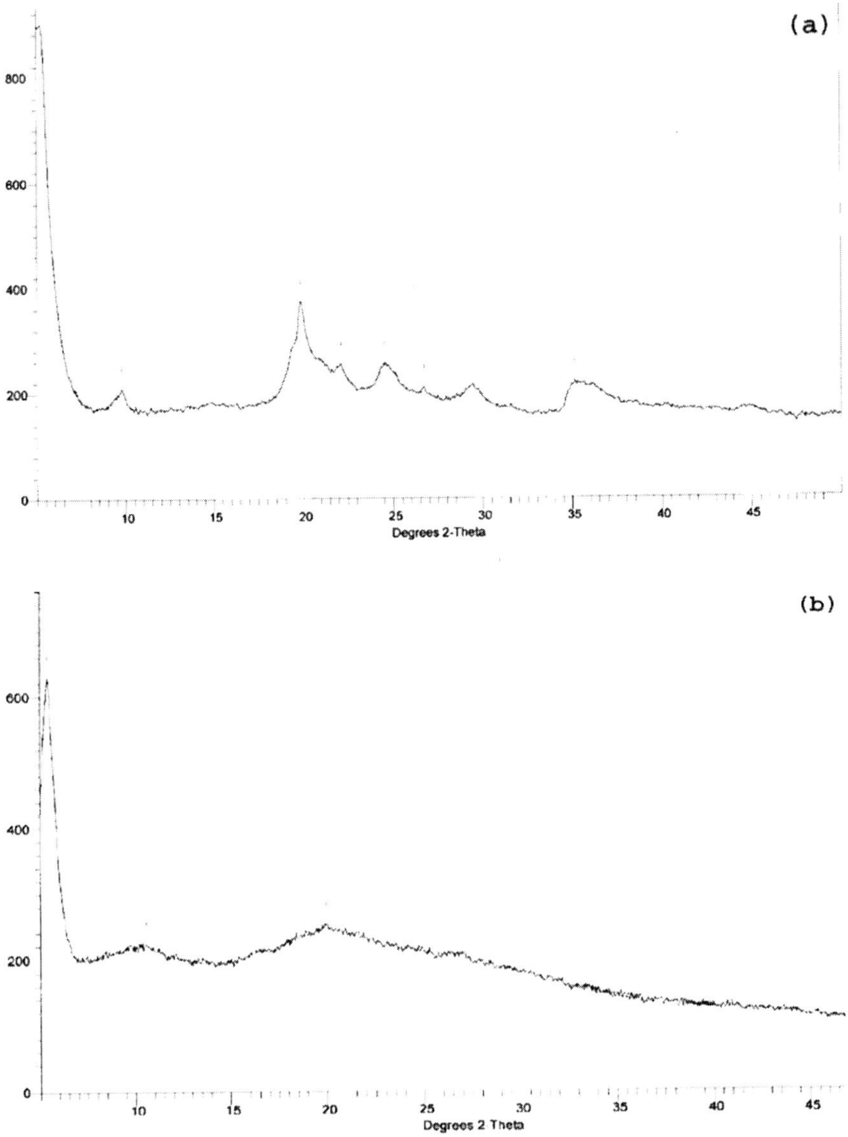

Figure 2. X-ray diffraction analysis showing the diffraction patterns that result from (a) Cloisite®30B in powder form, (b) 3 wt% sample of Cloisite®30B after dispersion in polyurethane.

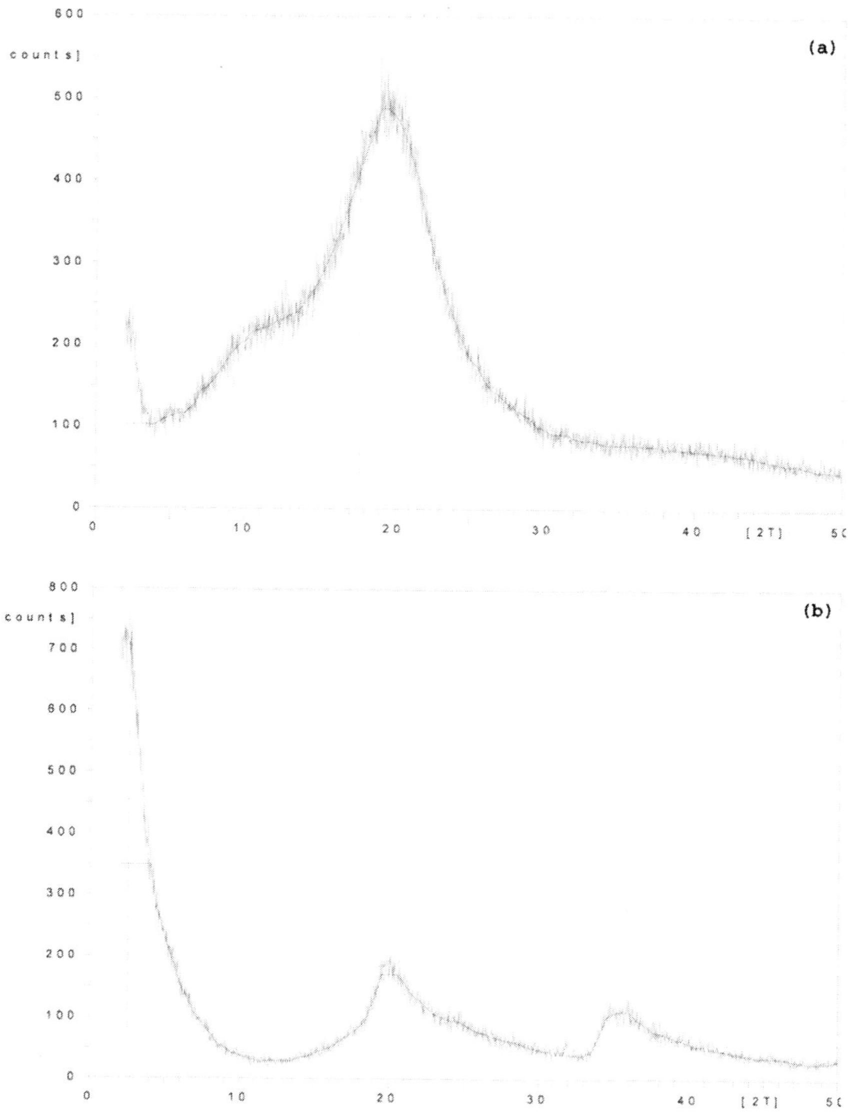

Figure 3. X-ray diffraction analysis showing the diffraction patterns that result from (a) organically modified synthetic layered silicates, (b) 3 wt% sample of processed synthetic layered silicates.

2Θ = 19 corresponding to 4.4 Å interlayer distance 2Θ = 10 corresponding to 8.6 Å interlayer distance 2Θ = 5 corresponding to 16.5 Å interlayer distance Similar to the situation for non-modified sample

(Cloisite® Na^+), the first three peaks of powder Cloisite® 30B have disappeared and shifted to 2Θ = 19 while the main peak with the highest intensity is still at the 2Θ = 5 corresponding peak.

Further observation of Cloisite® 30B behavior before and after process reveals an interesting occurrence. The overall intensity of peaks is slightly decreased. This fact is perhaps related to the use of ultrasonic deagglomerator for the dispersion process which in low viscosity systems may drive out the modifier through layered silicate layers.

Figure 3(a) shows the XRD curve of the synthetic nano layered silicate powder and **Figure 3**(b) shows the X-ray diffraction curve of polyurethane/3wt% synthetic nano layered silicate composite free film.

Synthetic nano layered silicate powder showed three main peaks as below:

d-values for powder:

2Θ = 35.5 corresponding to 2.5 Å interlayer distance 2Θ = 20 corresponding to 4.4 Å interlayer distance 2Θ = 3 corresponding to 30 Å interlayer distance Among three peaks detected by XRD, the 2Θ = 3 is the dominant peak while 35.5° and 20° peak's low intensity represents a relatively good modification operation.

d-values for processed synthetic nano layered silicate:

2Θ = 19 corresponding to 4.4 Å interlayer distance;

2Θ = 8 corresponding to 11 Å interlayer distance;

2Θ ≤ 2 that indicates the interlayer distance higher than 45 Å.

Similar to all previous cases, the 2Θ = 35.5 peak disappeared. Furthermore, the 2Θ = 19 peak is significantly amplified. Although the curve pattern in **Figure 3**(b) predicts a dramatic rise in curve height through 2Θ values lower than 5° unfortunately due to uncertainly of XRD results for 2Θ values lower than 2°, the real intensity cannot be detected.

The drying time recorder, pendulum hardness (Persoz), mandrel bending, Impact, Pull-off and gloss retention tests results are listed in **Table 1**. The initial gloss of all samples did not change noticeably unless for the 5 wt% sample of synthetic nano layered silicate. The initial gloss reduction of mineral clays can be interpreted as a result of their bigger size in comparison to synthetic silicate used in this study. The equality of samples filled with Cloisite® Na^+ and Cloisite® 30B shows that the modifier plays an insignificant role regarding initial gloss and the most important parameter to be taken into account is the primary size of the layered silicate.

Table 1. Drying time recorder, pendulum hardness (Persoz), mandrel bending, impact, pull-off and gloss retention test results.

Property	Blank PU	1 wt% synthetic nano	3 wt% synthetic nano	5 wt% synthetic nano	3 wt% Na^+ Clay	3 wt% 30B Clay
Gloss						
After 1 week	140	140	140	138	138	138
Drying time (min)						
Surface dry	50	55	55	60	58	57
Dry through	295	340	330	340	345	345
Pendulum hardness						
After 48 hrs	174	200	250	226	230	248
After 1 week	311	312	312	305	311	312
Gloss retention @ 60 degree						
Initial	137	137	140	-	138	138
After 70 double rub	100	107	113	-	102	107
Retention percentage	72%	78%	81%		66%	76%
Mandrel bending						
After 1 week	4 mm	4 mm	4 mm		4 mm	4 mm
Impact						
After 1 week	75 lb.in	75 lb.in	75 lb.in	-	75 lb.in	76 lb.in
Pull-off						
After 1 week	369 psi	442 psi	375 psi	256 psi	355 psi	400 psi

The gloss retention results are very interesting as we have 9% more gloss retention for synthetic layered silicate with a mere loading of 3 wt% nano filler. In case of Cloisite® Na^+, gloss retention result was even worse than blank polyurethane. It implies that poor interaction between Cloisite® Na+ and polymer matrix has led to noticeable a cohesion loss throughout the coating film in this sample. The sample containing

Cloisite® 30B shows fairly better results in comparison to blank polyurethane whereby its gloss retention is still 5% less than synthetic sample.

The impact and Mandrel tests did not show any significant difference. All samples could pass the minimum requirements successfully. The pendulum hardness test showed significant increase in hardness 48 hrs after application in all samples where 3 wt% synthetic nano layered silicate stood on top with a distinguishable value. This indicates that the 3 wt% synthetic nano layered silicate has been able to present a higher level of uniform dispersion in comparison to the other samples. Clearly, this result is only limited to the applied thickness in this study and cannot be generalized for other coatings with different thicknesses. Nevertheless, the coating hardness increase during the first 48hrs after application can remarkably help protect the clear coat against scratch. Normally in after sale market, automobile is released less than 48 hrs after repair while a complete curing occurs 7 days after application and thus clear coat is highly sensitive to scratch in the first few days. On the other hand, this immediate coating hardness can also minimize the depth of scratches that may occur during first days.

The final pendulum hardness (after 7 days) was almost the same for all samples. The reflectance spectra of samples containing 3 wt% synthetic nano layered silicate, 3 wt% Cloisite® Na^+, 3 wt% Cloisite® 30B and blank polyurethane (coated on a typical white solid paint) at radiation angles of 20°, 45°, 75° and 110° are shown in **Figure 4**.

In addition to the goniospectrophotometer L, a, b data, in order to find out whether the changes in reflectance spectra are detectable by human eye or not, ΔE was calculated. Pure L, a, b data of solid white paint with a blank polyurethane are listed in **Table 2**. For nano-filled samples, the results are listed in 4 different angles in Table 3. Colorimetric deviation averages are as below:

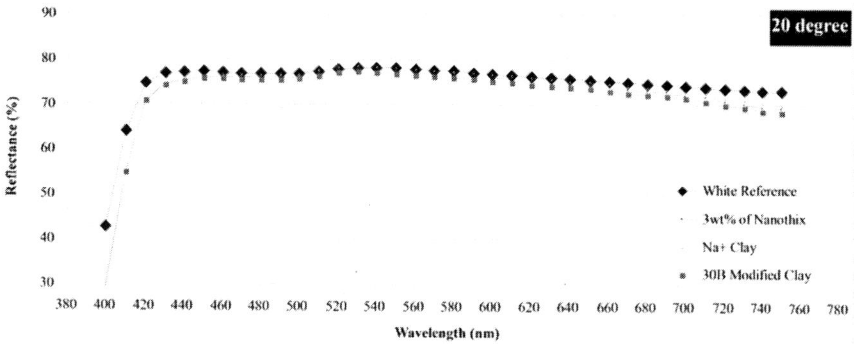
20 degree
Reflectance (%)
90
80
70
60
50
40
30
380 400 420 440 460 480 500 520 540 560 580 600 620 640 660 680 700 720 740 760 780
Wavelength (nm)
White Reference
3wt% of Nanothix
Na+ Clay
30B Modified Clay

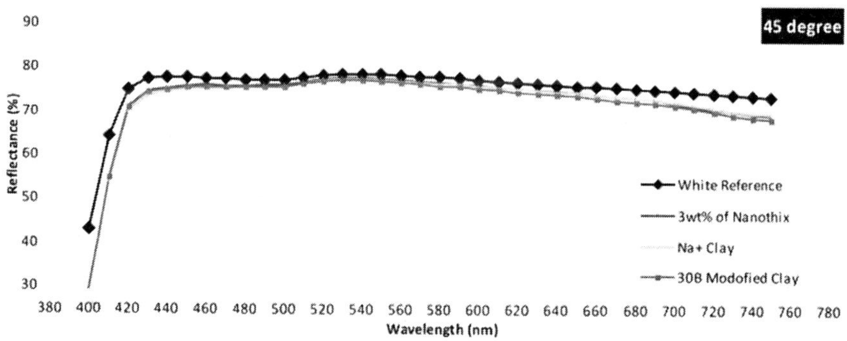
45 degree
Reflectance (%)
90
80
70
60
50
40
30
380 400 420 440 460 480 500 520 540 560 580 600 620 640 660 680 700 720 740 760 780
Wavelength (nm)
White Reference
3wt% of Nanothix
Na+ Clay
30B Modofied Clay

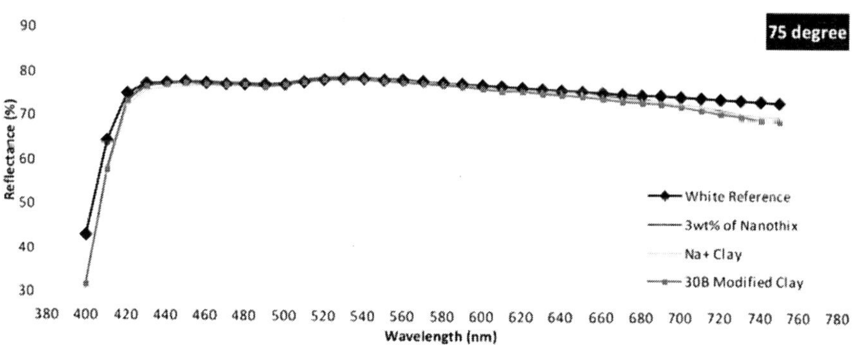
75 degree
Reflectance (%)
90
80
70
60
50
40
30
380 400 420 440 460 480 500 520 540 560 580 600 620 640 660 680 700 720 740 760 780
Wavelength (nm)
White Reference
3wt% of Nanothix
Na+ Clay
30B Modified Clay

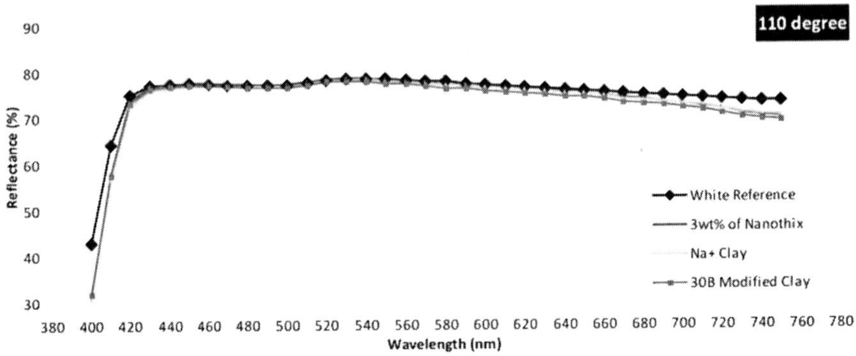

Figure 4. Reflectance curves of solid white paint coated substrate with blank clear coat, Cloisite® Na^+ filled clear coat, Cloisite®30B filled clear coat and nanothix filled clear coat in four different angle (under D65 light source). All samples contain 3 wt% of filler.

Synthetic Nanothix vs blank polyurethane

A: 0.66 D65: 0.91 TL84: 0.84 Mineral clay Cloisite® Na^+ vs blank polyurethane A: 0.87 D65: 1.2 TL84: 1.23 Mineral clay Cloisite® 30B vs blank polyurethane A: 0.81 D65: 0.94 TL84: 0.94 Among the three evaluated samples, synthetic nano layered silicate showed fairly better results while organically modified nano clay (Cloisite® 30B) also yielded acceptable results. Disregarding the slight shape differences between clay tactoids and an ideal sphere, the difference between Cloisite® 30B and Cloisite® Na^+ can be interpreted via Rayleigh equation as a result of low dispersion status which is in accordance with the XRD results. According to Rayleigh's equation, for discrete spherical particles in an amorphous polymer matrix, the reduction in light intensity due to scattering is defined as:

Table 2. Pure L, a, b data of solid white paint coated with blank PU for different light sources.

Solid white coated with blank polyurethane			
Illuminant	L*	a*	b*
A	90.462	−0.906	0.546
D65	90.518	−1.055	0.874
TL83	90.522	−0.914	0.756

Table 3. L, a, b and ΔE data of solid white paint coated with (a) 3 wt% nanothix, (b) 3 wt% Cloisite® 30B, (c) 3 wt% Cloisite® Na^{+}.

Sample	Pass/Fail	Illuminant		L*	a*	b*	ΔE
		A	20°	90.114	−1.288	1.252	0.87491
			45°	89.906	−1.276	1.191	0.92847
			75°	90.466	−1.309	0.631	0.41189
			110°	90.341	−1.318	0.434	0.44377
		D65	20°	90.185	−1.773	1.821	1.23419
Synthetic nano			45°	89.977	−1.732	1.747	1.23010
	Pass		75°	90.563	−1.599	1.16	0.61624
			110°	90.45	−1.589	0.975	0.54771
		TL83	20°	90.195	−1.328	1.925	1.28253
			45°	89.977	−1.319	1.782	1.23034
			75°	90.513	−1.279	1.002	0.44025
			110°	90.401	−1.3	0.84	0.41315
		A	20°	90.445	−1.282	1.48	1.00699
			45°	89.822	−1.196	1.634	1.29516
			75°	90.303	−1.214	1.083	0.63915
			110°	90.234	−1.219	0.901	0.52534
		D65	20°	90.507	−1.837	2.057	1.41814
Clay Na^{+}			45°	89.867	−1.767	2.181	1.62450
	Pass		75°	90.372	−1.632	1.606	0.94343
			110°	90.314	−1.618	1.434	0.81987
		TL83	20°	90.526	−1.351	2.229	1.53646
			45°	89.897	−1.302	2.323	1.73109
			75°	90.353	−1.244	1.558	0.88355
			110°	90.299	−1.261	1.423	0.78424
		A	20°	89.825	−1.401	0.945	0.90000
			45°	89.61	−1.363	1.012	1.07327
			75°	90.193	−1.385	0.447	0.55821
			110°	89.988	−1.374	0.276	0.71875
		D65	20°	89.92	−1.812	1.532	1.16774
Clay 30B			45°	89.698	−1.774	1.583	1.30079
	Pass		75°	90.306	−1.632	0.991	0.62575
			110°	90.11	−1.603	0.825	0.68496
		TL83	20°	89.9	−1.401	1.564	1.13001
			45°	89.684	−1.39	1.585	1.27124
			75°	90.239	−1.323	0.796	0.49897
			110°	90.043	−1.322	0.653	0.63758

$$I/I_o = e - \left(\frac{3V_p x r^3}{4\lambda^4} \left(\frac{n_p}{n_m - 1} \right) \right)$$

where I_o is the incident intensity, x is the optical path length, r is the particle radius, n_p is the reflective index of the particles, n_m is the reflective index of the matrixV is the volume fraction of particles and λ is the wavelength of light.

Scratch resistance test results showed an obvious improvement in scratch resistance of nano-filled clear coats in presence of nano filler. **Figure 5** shows the microscopic images of automatic scratch test for blank polyurethane, the 3 wt% synthetic nano layered silicate and 3 wt% Cloisite® 30B nano clay samples. In all cases, loaded weight is 2000 grams. **Figure 6** shows images of an incision made over polymer films.

Figure 7 shows microscopic images of different samples after 70 double rubs of abrasive which confirms the automatic scratch resistance test results. Some gloss retention results obtained from these samples are listed in **Table 1**. In addition, samples did not show noticeable difference in case of the chipping test at the applied thickness.

In order to trace the adhesion behavior of coatings, digital Pull-off test was employed. Results are listed in **Table 1**. In comparison with control sample (blank polyurethane), the 1 wt% nanothix filled sample and 3 wt% Cloisite® 30B filled polyurethane showed adhesion improvement. There was also a slight increase in adhesion for 3 wt% nanothix filled sample. Considering the operation method of pull-off, this adhesion increase is mainly related to the polymer film cohesion strength resulting from polymer/nano layers interaction. Comparing to blank polyurethane, the 5 wt% nano-filled sample showed significant decrease in adhesion equal to 30%.

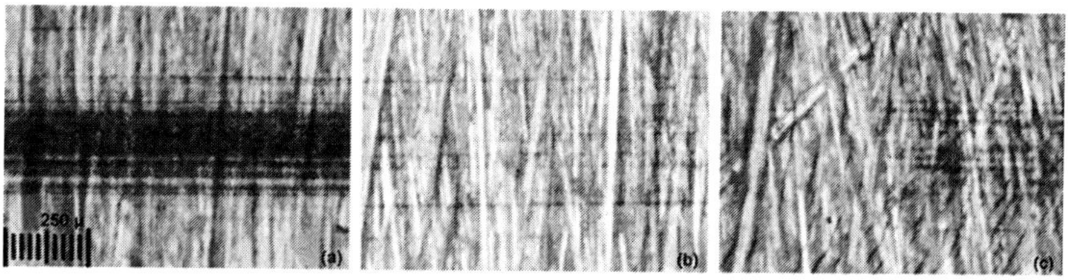

Figure 5. Microscopic images of (a) blank clear coat, (b) 3 wt% nanothix filled polyurethane, (c) 3 wt% nano Cloisite® 30B filled polyurethane. In all cases weight load is 2000gr.

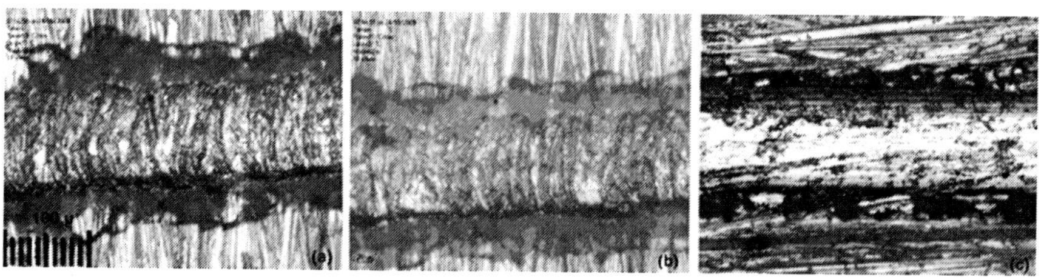

Figure 6. Microscopic images of (a) blank clear coat, (b) 3 wt% Cloisite® 30B filled polyurethane and (c) 3 wt% nanothix filled polyurethane.

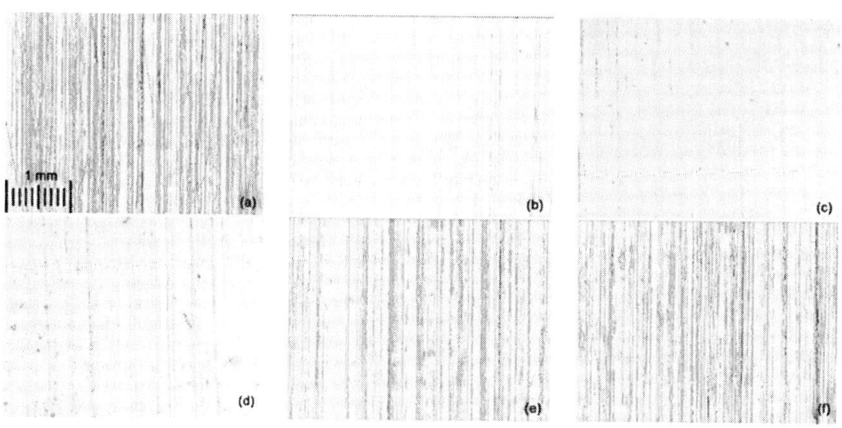

Figure 7. Microscopic images of (a) blank Polyurethane (b) 3 wt% nanothix filled polyurethane, (c) 3 wt% nano Cloisite® 30B filled polyurethane, (d) 3 wt% nano Cloisite® Na^+ filled polyurethane, (e) 1 wt% nanothix filled polyurethane and (f) 1 wt% nano Cloisite® 30B filled polyurethane after 70 double rubs.

4. CONCLUSIONS

Three different samples containing mineral and synthetic nano layered silicate were prepared and compared with a blank polyurethane clear coat. Among tested samples, 3 wt% sample of synthetic nano layered silicate yielded better results in most cases.

Scratch test showed an improvement after 7 days while a significant hardness increase was observed during merely 48 hours after coating application. This makes coating less sensitive to scratches that may occur during the first days after repair. Furthermore, the optical property of samples slightly changed where synthetic layered silicate contained sample presented least deviation. L, a, b and ΔE results showed that the deviation of the aforementioned sample is not detectable by the naked eye. Gloss retention results showed a significant improvement in comparison to blank polyurethane while initial gloss and dry to touch property of samples did not change noticeably. Adhesion presented different behavior in each case. It decreased in case of non-modified clay and 5 wt% of synthetic layered silicate while increased in case of modified clay and 1 wt% synthetic layered silicate. Adhesion did not alter noticeably for the 3 wt% synthetic sample. Other common requirements of an automotive clear coat such as bending resistance (flexibility) and impact resistance were not affected by nano filler presence at the applied thickness.

5. ACKNOWLEDGEMENTS

The authors gratefully acknowledge the support of Khosh Paint & Resin Co. Ltd.

REFERENCES

1. B. V. Gregorovich, K. Adamsons and L. Lin, "Scratch and Mar and Other Mechanical Properties as a Function of Chemical Structure for Automotive Refinish Coatings," Progress in Organic Coatings, Vol. 43, No. 1-3, 2001, pp. 175-187.

2. L. Lin, G. S. Blackman and R. R. Matheson, "Quantitative Characterization of Scratch and Mar Behavior of Polymer Coatings," Materials Science and Engineering, Vol. 317, No. 1-2, 2001, pp. 163-170.

3. A. Krupicka, M. Johansson and A. Hult, "Use and Interpretation of Scratch Tests on Ductile Polymer Coatings," Progress in Organic Coatings, Vol. 46, No. 1, 2003, pp. 32-48.

4. M. Groenewolt, "Highly Scratch Resistant Coatings for Automotive Applications," Progress in Organic Coatings, Vol. 61, No. 2-4, 2008, pp. 106-109.

5. E. Barna, B. Bommer, J. Kursteiner, A. Vital, O. V. Trzebiatowski, W. Koch, B. Schmid and T. Graule, "Innovative, Scratch Proof Nanocomposites for Clear Coatings," Composites Part A, Vol. 36, No. 4, 2005, pp. 473-480

6. Y. Kojima, A. Usuki, M. Kawasumi, A. Okada, Y. Fukushima, T. Kurauchi and O. Kamigaito, "Mechanical Properties of Nylon 6-Clay Hybrid," Journal of Materials Research, Vol. 8, No. 5, 1993, pp. 1185-1189.

7. A. Okada and A. Usuki, "The Chemistry of Polymer-Clay Hybrids," Materials Science and Engineering: C, Vol. 3, No. 2, 1995, pp. 109-115.

8. P. Messersmith and E. P. Giannelis, "Synthesis and Characterization of Layered Silicate-Epoxy Nanocomposites," Chemistry of Materials, Vol. 6, No. 10, 1994, pp. 1719-1725.

9. S. Manne and I. A. Aksay, "Thin Films and Nanolaminates Incorporating Organic/Inorganic Interfaces," Current Opinion in Solid State & Materials Science, Vol. 2, No. 3, 1997, pp. 358-364.

10. T, Lan and T. J. Pinnavaia, "Clay-Reinforced Epoxy Nanocomposites," Chemistry of Materials, Vol. 6, No. 12, 1994, pp. 2216-2219.

11. H. Shi, T. Lan and T. J. Pinnavaia, "Interfacial Effects on the Reinforcement Properties of Polymer-Organoclay Nanocomposites," Chemistry of Materials, Vol. 8, No. 8, 1996, pp. 1584-1587.

12. B. Ahmadi, S. M. Kassiriha, K. Khodabakhshi and E. R. Mafi, "Effect of Nano Layered Silicates on Automotive Polyurethane Refinish Clear Coat," Progress in Organic Coatings, Vol. 60, No. 2, 2007, pp. 99-104.

13. J. W. Gilman, "Flammability and Thermal Stability Studies of Polymer Layered-Silicate (Clay) Nanocomposites," Applied Clay Science, Vol. 15, No. 1-2, 1999, pp. 31-49.

14. F. Dietsche and R. Mulhaupt, "Thermal Properties and Flammability of Acrylic Nanocomposites Based Upon Organophilic Layered," Polymer Bulletin, Vol. 43, No. , 1999, pp. 395-402.

15. D. Porter, E. Metcalfe and M. J. K. Thomas, "Nanocomposite Fire Retardants—A Review," Fire Mater, Vol. 24, No. 1, 2000, pp. 45-52.

16. J. W. Gilman, C. L. Jackson, A. B. Morgan, R. Harris, E. Manias, E. P. Giannelis, et al., "Flammability Properties of Polymer–Layered-Silicate Nanocomposites. Polypropylene and Polystyrene Nanocomposites," Chemistry of Materials, Vol. 12, No. 7, 2000, pp. 1866-1873.

17. K. Yano, Y. Kojima, A. Usuki, A. Okada, T. Kurauchi and O. Kamigaito, "Synthesis and Properties of Polyimide-Clay Hybrid," Journal of Polymer Science Part A: Polymer Chemistry, Vol. 31, No. 10, 1993, pp. 2493- 2498.

18. T. Lan, P. D. Kaviratna and T. J. Pinnavaia, "On the Nature of Polyimide-Clay Hybrid Composites," Chemistry of Materials, Vol. 6, No. 5, 1994, pp. 573-575.

19. P. B. Messersmith and E. P. Giannelis, "Synthesis and Barrier Properties Of Poly(ϵ-caprolactone)-Layered Silicate

Nanocomposites," Journal of Polymer Science Part A: Polymer Chemistry, Vol. 33, No. 7, 1995, pp. 1047- 1057.

20. S. D. Burnside and E. P. Giannelis, "Synthesis and Properties of New Poly(dimethylsiloxane) Nanocomposites," Chemistry of Materials, Vol. 7, No. 9, 1995, pp. 1597- 1600.

21. M. S. Cho, H. J. Choi and K. To, "Effect of Ionic Pendent Groups on a Polyaniline-Based Electrorheological Fluid," Macromolecular Rapid Communications, Vol. 19, No. 6, 1998, pp. 271-273.

22. S. G. Kim, J. W. Kim, M. S. Cho, H. J. Choi and M. S. Jhon, "Viscoelastic Characterization of Semiconducting Dodecylbenzenesulfonic Acid Doped Polyaniline Electrorheological Suspensions," Journal of Applied Polymer Science, Vol. 79, No. 1, 2000, pp. 108-114.

23. J. M. Yeh, S. J. Liou, C. Y. Lin, C. Y. Cheng, Y. W. Chang and K. R. Lee, "Anticorrosively Enhanced PMMAClay Nanocomposite Materials with Quaternary Alkylphosphonium Salt as an Intercalating Agent," Chemistry of Materials, Vol. 14, No. 1, 2002, pp. 154-161.

24. J. M. Yeh, S. J. Liou, M. C. Lai, Y. W. Chang, C. Y. Huang, C. P. Chen, et al., "Comparative Studies of the Properties of Poly(methyl methacrylate)-Clay Nanocomposite Materials Prepared by in Situ Emulsion Polymerization and Solution Dispersion," Journal of Applied Polymer Science, Vol. 94, No. 5, 2004, pp. 1936-1946.

25. J. M. Yeh, S. J. Liou, H. J. Lu and H. Y. Huang, "Enhancement of Corrosion Protection Effect of Poly (Styrene-Co-Acrylonitrile) by the Incorporation of Nanolayers of Montmorillonite Clay into Copolymer Matrix," Journal of Applied Polymer Science, Vol. 92, No. 4, 2004, pp. 2269-2277.

26. J. M. Yeh, S. J. Liou, C. G. Lin, Y. P. Chang, Y. H. Yu and C. F. Cheng, "Effective Enhancement of Anticorrosive Properties of

Polystyrene by Polystyrene-Clay Nanocomposite Materials," Journal of Applied Polymer Science, Vol. 92, No. 3, 2004, pp. 1970-1976.

27. J. M. Yeh, C. L. Chen, Y. C. Chen, C. Y. Ma, H. Y. Huang, Y. H. Yu, "Enhanced Corrosion Prevention Effect of Polysulfone-Clay Nanocomposite Materials Prepared by Solution Dispersion," Journal of Applied Polymer Science, Vol. Vol. 92, No. 1, 2004, pp. 631-637.

28. J. M. Yeh, H. Y. Huang, C. L. Chen, W. F. Su and Y. H. Yu "Siloxane-Modified Epoxy Resin-Clay Nanocomposite Coatings with Advanced Anticorrosive Properties Prepared by a Solution Dispersion Approach," Surface and Coatings Technology, Vol. 200, No. 8, 2006, pp. 2753-2763.

29. Y. H. Yu, J. M. Yeh, S. J. Liou and Y. P. Chang, "Organo-Soluble Polyimide (TBAPP-OPDA)/Clay Nanocomposite Materials with Advanced Anticorrosive Properties Prepared from Solution Dispersion Technique," Acta Materialia, Vol. 52, No. 2, 2004, pp. 475-486.

30. J. M. Yeh, C. F. Hsieh, J. H. Jaw, T. H. Kuo, H. Y. Huang, C. L. Lin, et al., "Organo-Soluble Polyimde (ODA-BSAA)/Montmorillonite Nanocomposite Materials Prepared by Solution Dispersion Technique," Journal of Applied Polymer Science, Vol. 95, No. 5, 2005, pp. 1082- 1090.

31. Y. H. Yu, J. M. Yeh, S. J. Liou, C. L. Chen, D. J. Liaw and H. Y. Lu, "Preparation and Properties of PolyimideClay Nanocomposite Materials for Anticorrosion Application," Journal of Applied Polymer Science, Vol. 92, No. 6, 2004, pp. 3573-3582.

32. J. M. Yeh, C. L. Chen, T. H. Kuo, W. F. Su, H. Y. Huang, D. J. Liaw, et al., "Preparation and Properties of (BATBODPA) Polyimide-Clay Nanocomposite Materials," Journal of Applied Polymer Science, Vol. 92, No. 2, 2004, pp. 1072-1079.

33. J. M. Yeh, S. J. Liou, C. Y. Lai, P. C. Wu, T. Y. Tsai, "Enhancement of Corrosion Protection Effect in Polyaniline via the Formation of Polyaniline-Clay Nanocomposite Materials," Chemistry of Materials, Vol. 13, No. 3, 2001, pp. 1131-1136.

34. J. M. Yeh, C. L. Chen, Y. C. Chen, C. Y. Ma, K. R. Lee, Y. Wei, et al., "Enhancement of Corrosion Protection Effect of Poly(o-ethoxyaniline) via the Formation of Poly(oethoxyaniline)-Clay Nanocomposite Materials," Polymer, Vol. 43, No. 9, 2002, pp. 2729-2736.

35. K. C. Chang, M. C. Lai, C. W. Peng, Y. T. Chen, J. M. Yeh, C. L. Lin, et al., "Comparative Studies on the Corrosion Protection Effect of DBSA-Doped Polyaniline Prepared from in Situ Emulsion Polymerization in the Presence of Hydrophilic Na^{+}-MMT and Organophilic OrganoMMT Clay Platelets," Electrochimica Acta, Vol. 51, No. 26, 2006, pp. 5645-5653.

36. Y. H. Yu, C. C. Jen, H. Y. Huang, P. C. Wu, C. C. Huang and J. M. Yeh, "Preparation and Properties of Heterocyclically Conjugated Poly(3-hexylthiophene)-Clay Nanocomposite Materials," Journal of Applied Polymer Science, Vol. 91, No. 6, 2004, pp. 3438-3446.

37. J. M. Yeh, C. P. Chin and S. Chang, "Enhanced Corrosion Protection Coatings Prepared from Soluble Electronically Conductive Polypyrrole-Clay Nanocomposite Materials," Journal of Applied Polymer Science, Vol. 88, No. 14, 2003, pp. 3264-3272.

38. J. M. Yeh and C. P. Chin, "Structure and Properties of Poly(o-methoxyaniline)-Clay Nanocomposite Materials," Journal of Applied Polymer Science, Vol. 88, No. 4, 2003, pp. 1072-1080.

39. R. E. Grim, "Clay Mineralogy," second ed., New York: McGraw Hill, 1968, p.362.

40. R. Tantatherdtam, "Reinforcement of Natural Rubber Latex by Nanosize Montmorillonite Clay," Ph.D. Thesis, Department of Materials Science and Engineering, University Park, 2003.

41. E. P. Giannelis, "Polymer Layered Silicate NanocompoSites," Advanced Materials, Vol. 8, No. 1, 1996, pp. 29- 35.

42. P. C. LeBaron, Z. Wang and T. J. Pinnavaia, "PolymerLayered Silicate Nanocomposites: An Overview," Applied Clay Science, Vol. 15, No. 1-2, 1999, pp. 11-29.

43. Y. Fukushima and S. Inagaki, "Synthesis of an Intercalated Compound of Montmorillonite and 6-Polyamide," Journal of Inclusion Phenomena and Macrocyclic Chemistry, Vol. 5, No. 4, 1987, pp. 473-482.

44. R. S. Berns, "Billmeyer and Saltzman's Principles of Color Technology," Third Edition, Wiley, New York, 2000, p. 127.

Chapter 7

Modeling Radiation-Induced Degradation in Top-Gated Epitaxial Graphene Field-Effect-Transistors (FETs)

Ivan S. Esqueda [1,*], Cory D. Cress [2], Travis J. Anderson [2], Jonathan R. Ahlbin [1], Michael Bajura [1], Michael Fritze [1] and Jeong-S. Moon [3]

[1]*Information Sciences Institute, University of Southern California, Arlington, VA 22203, USA*

[2]*U.S. Naval Research Laboratory, Electronics Science and Technology Division, Washington, DC 20375, USA*

[3]*HRL Laboratories LLC, Malibu, CA 90265, USA*

ABSTRACT

This paper investigates total ionizing dose (TID) effects in top-gated epitaxial graphene field-effect-transistors (GFETs). Measurements reveal voltage shifts in the current-voltage (*I-V*) characteristics and degradation of carrier mobility and minimum conductivity, consistent with the buildup of oxide-trapped charges. A semi-empirical approach for modeling radiation-induced degradation in GFETs effective carrier mobility is described in the paper. The modeling approach describes Coulomb and short-range scattering based on calculations of charge and effective vertical field that incorporate radiation-induced oxide trapped charges. The transition from the dominant scattering mechanism is correctly described as a function of effective field and oxide trapped charge density. Comparison with experimental data results in good qualitative agreement when including an empirical component to account for scatterer transparency in the low field regime.

Keywords: graphene; field-effect-transistors (FETs); total ionizing dose (TID); radiation; conductivity; mobility

1. INTRODUCTION

Graphene field effect transistors (GFETs) have received significant interest from the electron-device community in the past few years due to the novel electronic properties of graphene, such as ambipolar field effect, high carrier mobilities and high thermal conductivity [1,2,3]. Due to the lack of a band-gap, GFETs are unsuitable for logic applications. However, the excellent transport properties described above make graphene FETs a good choice for high-speed RF applications where a high I_{on}/I_{off} ratio is not required. The work presented here investigates the total dose response of RF GFETs intended for space and other applications that encounter ionizing radiation environments and describes extraction and modeling techniques for radiation induced damage.

As described in [4], there have been a few studies of ionizing radiation effects on graphene and on graphene-based materials. Initial investigations of total ionizing dose (TID) effects on graphene-based devices were done using back-gated GFETs and performed under different conditions (e.g., irradiated and characterized under vacuum or in air) resulting in a different radiation response [4,5,6]. In a typical back-gated GFET test structure the graphene layer is exposed causing the radiation response to be strongly dependent on the experimental ambient condition. As described in [5], different mechanisms can contribute to the radiation response. For example, in

[4,7] it was reported that chemical doping (*p*-type) that results from oxygen adsorption and/or reactions with the graphene layer during X-ray exposure resulted in positive shifts in the current-voltage (*I-V*) characteristics of devices irradiated in air. In [5], irradiation of back-gated GFETs in a controlled environment (*i.e.*, under static vacuum) confirmed that GFETs are susceptible to oxide charge trapping similar to silicon (Si) metal-oxide-semiconductor FETs (MOSFETs) as manifested by negative voltage shifts in the *I-V* characteristics. Thus, carefully controlling the experimental environment when investigating the basic radiation mechanisms in back-gated GFETs with exposed graphene layers is crucial [5]. Similar results have been reported for the radiation response of single-walled carbon nanotube thin-film-transistors [8].

As explained in [2], back-gated GFETs are appropriate for proof-of-concept purposes, but cannot be easily integrated with other components. Moreover, the thick dielectric layers in back-gated GFETs are highly vulnerable to radiation-induced charge buildup. As described in [9], practical graphene transistors must be patterned with local back-gates or otherwise have individual top-gate electrodes. In this work, we present evidence of TID effects in top-gated epitaxial GFETs. Oxide trapped charge near the SiO_2/graphene channel interface is identified as the main contributing mechanism to radiation-induced degradation. The buildup of oxide-trapped charges results in negative voltage shifts in the *I-V* characteristics and degradation in carrier mobility. These effects are determined experimentally by *in situ* characterization of GFETs exposed to Co-60 gamma rays. Extractions of radiation-induced oxide-trapped charge density (N_{ot}) are incorporated into a semi-empirical model for GFET effective carrier mobility. The modeling approach is verified through the comparison with GFET effective electron mobility and its radiation response extracted from the degraded *I-V* characteristics.

2. EXPERIMENTAL SECTION

Top-gated GFETs fabricated with epitaxial graphene layers grown on Si-face 6H-SiC substrates via Si sublimation were used in this study [10]. A schematic diagram of the top-gated GFET and a scanning electron microscope image of the test structure are shown in Figure 1a,b [10]. The top-gated GFETs were fabricated using Ti/Pt/Au source and drain contacts as well as Ti/Pt/Au metal gates over a 35 nm SiO_2 dielectric layer deposited via electron beam evaporation [10,11]. Two-finger GFET structures with a 3 µm gate length (*L*) and various gate widths (*W*) were selected for total dose experiments. The test structures were exposed to Co-60 gamma rays at the Naval Research Laboratory (NRL) at a dose rate of ~910 rad(Si)/s with a positive gate bias of V_g = 3 V and all other terminals grounded. Electrical characterization was performed following step-stress irradiations up to a total dose of 1000 krad(Si). The characterization consisted of measuring the output characteristics (*i.e.*, I_d-V_{ds}) by sweeping V_d from 0 to 3 V and stepping V_g from 2 V to −4 V in −1.0 V steps for V_s = 0, and the transfer characteristics (*i.e.*, I_d-V_{gs}) for V_d = 0.1 and 1 V for V_s = 0. For the transfer characteristics dual gate sweeps were done with V_g being swept from 2 V to −5 V then back to 2 V to observe the effects of gate hysteresis. In the top-gated GFET test structures used in this study, the graphene layer was not directly exposed to the experimental environment (*i.e.*, it is covered by the SiO_2/metal-gate stack; see Figure 1). Nevertheless, half of the samples were irradiated in air and the other half with a continuous flow of ultra pure N_2 purging oxygen, moisture and other contaminants from the irradiation vessel. Both cases revealed negative shifts in the *I*-*V* characteristics but significant shifts in the minimum conductivity were only observed for devices tested in air. Additionally, gate hysteresis did not change significantly as a function of TID but consistent shifts in the negative and positive $V_g$sweeps were observed in the post-irradiation I_d-V_{gs} characteristics of all devices. These shifts are due to radiation-induced positive oxide trapped charge and were analyzed independently of the mechanisms that cause hysteresis. The results are summarized and discussed in Section 3.

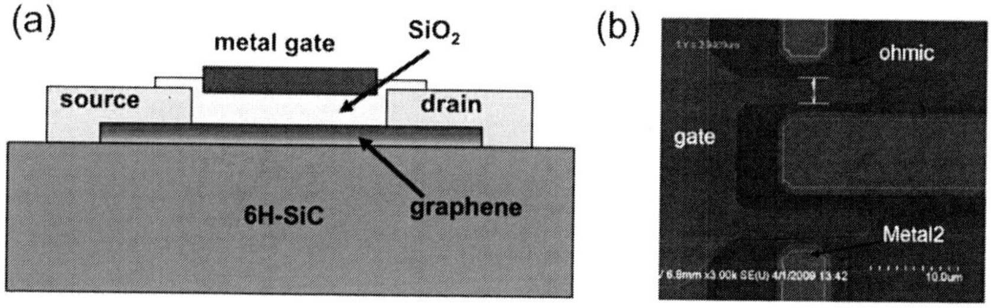

Figure 1. (**a**) Schematic diagram of top-gated graphene FET (GFET) fabricated with epitaxial graphene layers grown on Si-face 6H-SiC substrates via Si sublimation; (**b**) Scanning electron microscope (SEM) image of the GFET test structure.

3. RESULTS AND DISCUSSION

3.1. Current-Voltage Characteristics

Shown in Figure 2a are the I_d-V_g characteristics for a two-finger GFET with W = 25 µm and L = 3 µm following step-stress irradiations in air up to 1000 krad(Si) and measured using V_{ds} = 1 V. The measurements in Figure 2a reveal negative shifts in the I-V characteristics as well as an increase in the minimum drain current ($I_{d\text{-}min}$) as a function of TID. These results indicate a buildup of N_{ot} near the SiO_2/graphene layer interface shifting the minimum conductivity gate voltage (*i.e.*, the Dirac voltage V_{Dirac}) negatively as well as increasing the minimum conductivity (σ_{min}). The increased σ_{min} results from increasing graphene channel intrinsic charge density (q_{int}) as a function of dose (*i.e.*, with time during the experiment). In Figure 2a the solid lines are I_d measurements from positive to negative V_g sweeps and the dashed line are from negative to positive V_g sweep measurements as indicated by the arrows in the 0 krad(Si) curve. These are included to indicate hysteresis effects even prior to irradiation. Figure 2b is a plot of the extracted shifts in the Dirac point (*i.e.*, $\Delta V_{Dirac} = V_{Dirac} - V_{Dirac0}$)

and the relative increase in $I_{d\text{-}min}$ (*i.e.*, $I_{d\text{-}min}/I_{d\text{-}min0}$) from the data in Figure 2a. V_{Dirac0} and $I_{d\text{-}min0}$ are respectively the Dirac voltage and I_d at $V_g = V_{Dirac}$ for 0 krad(Si) (*i.e.*, prior to irradiation). The extractions in Figure 2b indicate a monotonic shift in V_{Dirac} and an increase in $I_{d\text{-}min}$ with total dose.

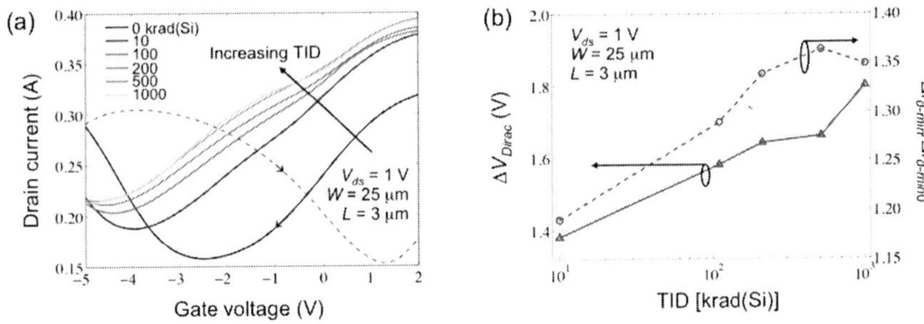

Figure 2. (**a**) Drain current (I_d) plotted as a function of the gate voltage (V_g) for increasing total ionizing dose (TID) up to 1000 krad(Si), V_d = 1 V, V_s = 0 V. Measurements are for a two-finger GFET with W = 25 µm and L = 3 µm; (**b**) Extracted shifts in the Dirac voltage (ΔV_{Dirac}) and relative increase in $I_{d\text{-}min}$plotted as a function of TID.

Shown in Figure 3a are the I_d-V_d characteristics for V_{gs} = 0 V, −1 V and −2 V measured at the same dose levels as show in Figure 2. The measurements in Figure 3 indicate an increase in I_d as a function of total dose for all values of V_{gs}. The increase in I_d is consistent with the negative shifts in I_d-V_g characteristics (*i.e.*, at a given $V_g > V_{Dirac}$, $I_d$will increase as a function of dose for all values of V_d due to the negative voltage shift in the I_d-V_g characteristics). Using the I_d-V_d measurements for all V_g steps (*i.e.*, from 2 V to −4 V), it is possible to represent I_d-V_g curves for any V_{ds} value contained within the sweep range of V_d (*i.e.*, between 0 and 3 V). For example, shown in Figure 3b are representations of I_d-V_{gs} for V_{ds} = 0.1 V and for all dose levels. In Figure 3b symbols are data and solid lines are a least squares fits to the data. Figure 4 is a plot of ΔV_{Dirac} and $\Delta\sigma_{min}$ obtained from the data in Figure 3b. By comparison, the I_d-V_g representations retain the TID effects observed in Figure 2.

Another TID effect, although not completely apparent from the measurements in Figure 2, Figure 3, is the degradation of carrier mobility. Degradation of carrier mobility is due to increased carrier scattering mechanisms that result from the buildup of oxide-trapped charges near the SiO_2/graphene channel interface [5,12]. The increase in carrier scattering as a function of ionizing radiation is demonstrated in section 3.3 through calculations of effective electron mobility (μ_{eff}) based on extractions of drain conductance $(g_{ds} = \partial I_d/\partial V_{ds})$, ΔV_{Dirac} and $\Delta\sigma_{min}$. Using ΔV_{Dirac} and $\Delta\sigma_{min}$ plotted in Figure 4 (extracted from the I_d-V_g representations inFigure 3b) ensures that any contributions from hysteresis effects are equivalent for the extractions of g_{ds} since the same I_d-V_{ds} measurements are used for all of the extractions.

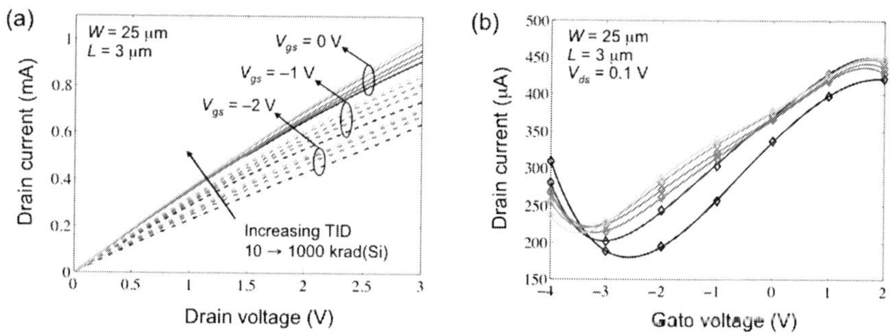

Figure 3. (**a**) Drain current (I_d) plotted as a function of the drain voltage (V_d) for increasing TID up to 1000 krad(Si), V_g = 0, −1 and −2 V, V_s = 0 V. Measurements are for a two-finger graphene field effect transistors (GFET) with W = 25 μm and L = 3 μm; (**b**) I_d-V_g representation for V_d = 0.1 V and for all dose levels. Symbols are data and solid lines are a least squares fits.

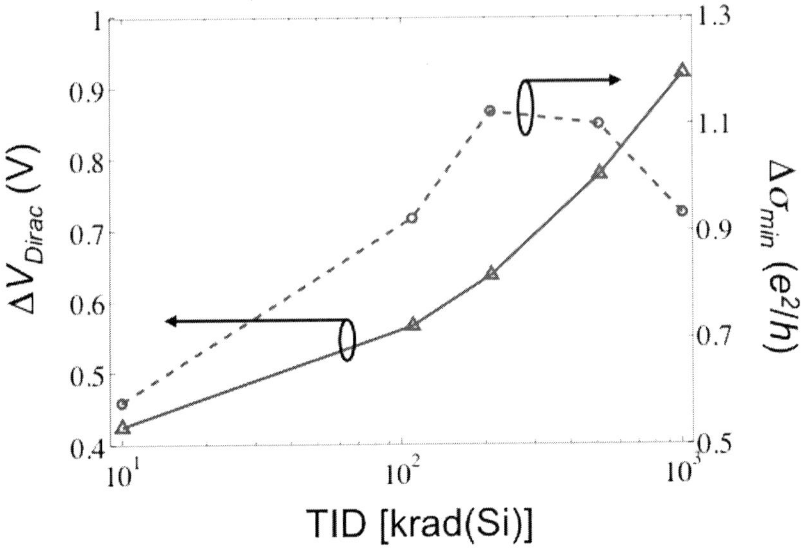

Figure 4. Shifts in the Dirac voltage ($\Delta V_{Dirac} = V_{Dirac} - V_{Dirac0}$) and minimum conductivity ($\Delta\sigma_{min} = \sigma_{min} - \sigma_{min0}$) obtained for V_{ds} = 0.1 V in Figure 3b. Note: V_{Dirac0} = −2.58 V, σ_{min0} = 4.7 e^2/h (e = 1.6 $\times 10^{19}$ C, h = 6.63 $\times 10^{-34}$ J-s).

3.2. Experimental Ambient Conditions

Shown in Figure 5a,b are the I_d-V_g and the I_d-V_d characteristics for a GFET device with W = 12 μm and L = 3 μm irradiated with a continuous flow of ultra pure N_2 purging the testing vessel. The response observed in Figure 5a,b reveals negative voltage shifts consistent with positive trapped charge as seen for devices irradiated in air. However, the devices irradiated with the N_2 purge have less hysteresis and show negligible shifts in σ_{min}. These differences indicate that molecular adsorption at the SiO_2 surface near the graphene channel (resulting from moisture and contaminants in the experimental environment) does have an effect, albeit reduced, on the electrical characteristics of top-gated GFET test structures. Therefore, it is possible that any increase in σ_{min} might be due to contaminants in the testing environment and not an effect of ionizing radiation exposure. In that case, the observed increase in σ_{min} for devices tested in air might be the result from accumulated adsorption of contaminants in the testing environment. For devices exposed with flowing N_2, the concentration of the contaminants was significantly reduced and therefore σ_{min} did not change significantly. Figure 6 plots extractions of hysteresis width (*h*) as a function of TID for both devices irradiated in air and with flowing N_2. Hysteresis width is extracted as the difference in the V_g intercept at a constant current for the negative and positive sweeps. The extractions of *h* reveal larger widths for devices tested in air, as expected due molecular adsorption at the surface of the SiO_2 contributing to charge trapping mechanisms that cause hysteresis [13]. However both experiments show negligible changes in hysteresis with TID. The radiation-induced degradation of effective carrier mobility is analyzed independently of the mechanisms that cause hysteresis in Section 3.3 based on these observations.

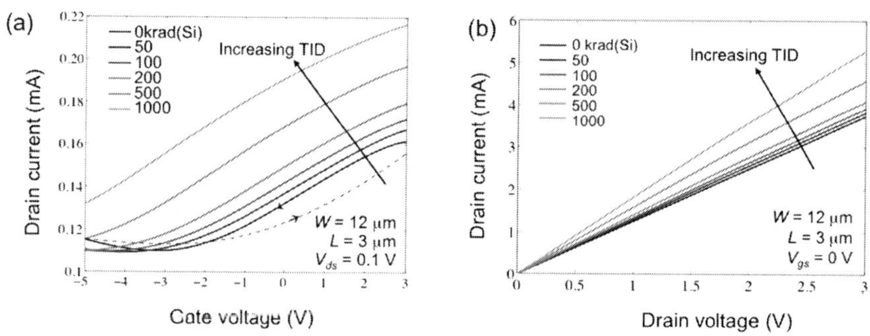

Figure 5. (**a**) Drain current (I_d) plotted as a function of the gate voltage (V_g) for increasing TID up to 1000 krad(Si), V_d = 1 V, V_s = 0 V. Measurements are for a two-finger GFET with W = 12 μm and L = 3 μm; (**b**) Drain current (I_d) plotted as a function of the drain voltage (V_d) for V_s = 0 V and for increasing TID up to 1000 krad(Si).

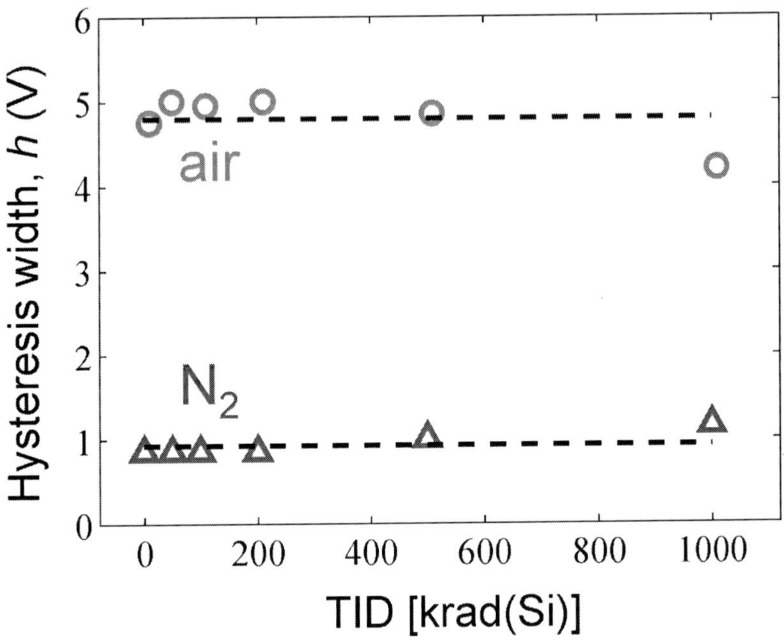

Figure 6. Extraction of hysteresis width (*h*) as a function of TID for devices tested in air and with flowing N_2 purge. Symbols are data and dashed line is the mean hysteresis width.

3.3. Mobility Degradation

The radiation-induced degradation of carrier mobility is determined through calculations of effective electron mobility in the graphene layer (μ_{eff}) as a function of the effective vertical field (E_{eff}). The effective electron mobility is given by

$$\mu_{eff} = \frac{(L/W)g_{ds}}{Q_{gc}}, \tag{1}$$

where Q_{gc} is the graphene channel charge density [14]. The effective vertical field in the graphene channel is estimated as [10]

$$E_{eff} = \frac{Q_{gc}}{\varepsilon_{ox}}, \tag{2}$$

and the graphene channel charge density is calculated as

$$Q_{gc} = C_{ox}\left(V_g - V_{Dirac}\right) + q_{int}. \tag{3}$$

Here, q_{int} is the graphene channel intrinsic charge density that results from charge inhomogeneity in the graphene layer typically described as electron-hole puddles that arise from the random charged impurity potential fluctuations and contribute to the non-zero minimum conductivity at the Dirac point [15,16,17,18].

Shown in Figure 7a is g_{ds} plotted as a function of V_g for all dose levels. Calculations of μ_{eff} using Equations 1–3 are plotted in Figure 7b as a function of E_{eff} and for increasing TID. In the calculations, q_{int} is estimated by σ_{min}/μ_{FE}, μ_{FE} is the peak field effect mobility and g_{ds} is extracted for V_{ds} = 0.1 V. The results in Figure 7b reveal a decrease in μ_{eff} with increasing total dose as can be observed at E_{eff} ~ 1 MV/cm. Deviations from the pre-irradiation $\mu_{eff}(E_{eff})$ characteristics are due to the increased contribution of scattering mechanisms that results from radiation-induced oxide trapped charges. The deviations from the pre-irradiation $\mu_{eff}(E_{eff})$ curve are expected to continue towards lower values of E_{eff} (*i.e.*, towards lower carrier densities $n \propto E_{eff}$) where scattering due to charged impurities is the dominant mechanism. Other scattering mechanisms such as short-range scattering dominate for large carrier densities (n) [15,19] (*i.e.*, for large E_{eff}) and the $\mu_{eff}(E_{eff})$ characteristics converge to a universal response. In Figure 7b this convergence appears for E_{eff} > 2 MV/cm.

In the approach presented here, the calculations of Q_{gc} (and E_{eff}) capture the contribution of radiation-induced oxide trapped charges through the shifts in V_{Dirac}. Therefore, this approach allows determining the radiation-induced degradation of carrier mobility in GFET devices without ambiguities introduced by voltage shifts in the I_d-V_g characteristics and the changes in σ_{min}. These ambiguities exist, for example, when plotting μ_{FE} as a function of V_g at various levels of TID. Nonetheless, it should be noted that discrepancies in the calculations of μ_{eff} could result from the approximations made in Equations 1–3. Interface states have not been included in the extractions presented here, but can be incorporated by including an interface trap charge density (bias-dependent) term $Q_{it}(V_g, N_{it})$ in Equation 3. Additionally, non-uniform buildup of N_{ot} may result from non-uniform field distributions in the gate oxide (e.g., in regions not covered by the top gate). These conditions would require a more sophisticated calculation of Q_{gc}, but the basic methodology presented here would still apply. The following section describes a semi-empirical modeling approach for the degradation of effective carrier mobility based on the experimental observations of $\mu_{eff}(E_{eff})$ and its response to ionizing radiation.

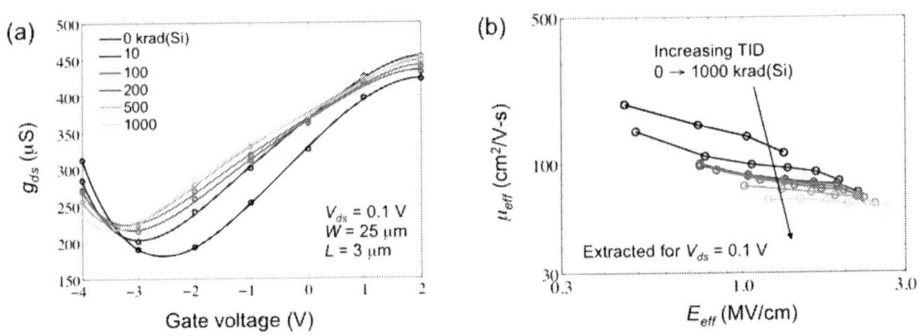

Figure 7. (**a**) Drain conductance (g_{ds}) plotted as a function of the gate voltage at V_d = 0.1 V for increasing dose levels up to 1000 krad(Si); (**b**) Effective electron mobility (μ_{eff}) plotted as a function of the effective vertical field (E_{eff}) for increasing total dose levels up to 1000 krad(Si).

4. MODELING

This section presents a semi-empirical modeling approach for describing effective carrier mobility in GFET devices and its degradation due ionizing radiation exposure. The radiation-induced shifts in V_{Dirac} can be directly related to N_{ot} as

$$\Delta V_{Dirac} = -q\Delta N_{ot}/C_{ox}. \tag{4}$$

In (4), N_{ot} accounts for the buildup of radiation-induced trapped charges that image charge into the graphene layer, C_{ox} is the oxide capacitance per unit area and q is the electronic charge. The effective electron carrier mobility is expressed using Matthiessen's rule as

$$\frac{1}{\mu_{eff}} = \frac{1}{\mu_C(N_{ot}, Q_{gc})} + \frac{1}{\mu_{SR}(E_{eff})}, \tag{5}$$

where E_{eff} is calculated using Equations 2 and 3 and V_{Dirac} as given by Equation 4 (*i.e.*, $V_{Dirac} = V_{Dirac0} - q\Delta N_{ot}/C_{ox}$). In Equation 5, μ_C is the Coulomb scattering limited effective carrier mobility and is modeled as a function of N_{ot} and Q_{gc} as

$$\mu_C^{-1} = \alpha(N_{ot}/N_0)\left(\frac{1}{1 + Q_{gc}/Q_0}\right)^{\beta}. \tag{6}$$

In Equation 5, μ_{SR} corresponds to the short-range scattering that dominates for large n and is given by

$$\mu_{SR}^{-1} = \mu_{SR0}^{-1}\left(E_{eff}/E_0\right)^{a}. \tag{7}$$

Equations 6 and 7 are adapted from a semi-empirical model for inversion layer mobility in Si MOSFETs described in [20]. Figure 8a shows calculations of μ_{eff} using Equations 5–7 and plotted as a function of E_{eff} for increasing N_{ot}. For the calculations in Figure 8a, α = 0.005 V-s/cm^2, μ_{SR0} = 1000 cm^2/V-s, β = 2, a = 0.75, $N_0 = 10^{12}$ cm^{-2}, $Q_0 = 10^{-6}$C/cm^2, $E_0 = 10^6$ V/cm, E_{eff} is in units of V/cm, Q_{gc} in units of C/cm^2 and N_{ot} is in units of cm^{-2}. The calculations inFigure 8a correctly describe the reduction in μ_{eff} with N_{ot} (*i.e.*, $\mu_{eff} \propto 1/N_{ot}$) for low n and the merging with μ_{SR}(dashed lined) for large n. In the model, parameters a, μ_{SR0}, α and β can be treated as fitting parameters and Q_0,E_0 and N_0 are constants that allow a simpler interpretation of the units for E_{eff}, Q_{gc} and N_{ot}.

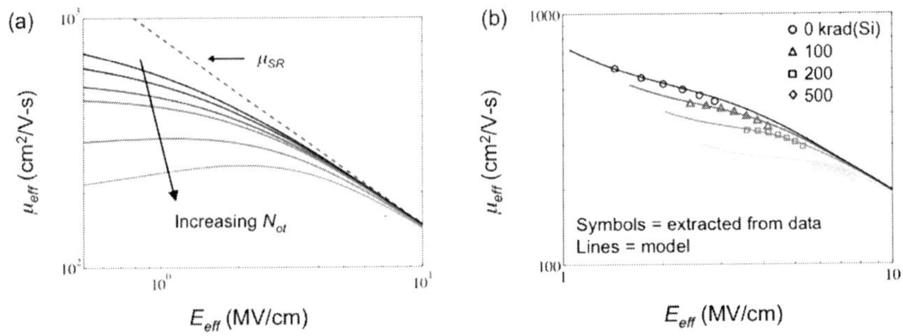

Figure 8. (**a**) Calculations of the effective electron mobility (μ_{eff}) using Equations 5–7 and plotted as a function of the effective vertical field (E_{eff}) for increasing values of N_{ot}. The dashed line corresponds to the short-range mobility given by Equation 7; (**b**) Comparison of μ_{eff} extracted from data inFigure 5 (symbols) and model calculations (lines) including scatterer transparency.

An increase in μ_{eff} for decreasing E_{eff} < ~1 MV/cm can be modeled empirically adding $\mu_0 = A(E_0/E_{eff})^b$ to the calculation of μ_{eff} given by Equation 5. This allows obtaining good agreements with the experimental results. This increase in mobility for decreasing n has been reported in [19] and attributed to scatterer transparency occurring when carrier wavelengths exceed the spacing between scatterers. In this work, the μ_{eff} dependence to scatterer transparency in the low n limit is modeled independently from charged impurity and short-range scattering using the aforementioned empirical form. Shown in Figure 8b are calculations of μ_{eff} compared with extractions from experimental data from irradiated GFETs. For the calculations in Figure 8b the fitting parameters are α = 0.08 V-s/cm^2, μ_{SR0} = 2000 cm^2/V-s, A = 340 cm^2/V-s, β = 4, a = 0.85, b = 1.08, $N_0 = 10^{12}$ cm^{-2}, $Q_0 = 10^{-6}$ C/cm^2 and $E_0 = 10^6$ V/cm. As shown in Figure 8b, the model calculations fit the experimental data with good qualitative agreement validating the presented approach. Additional experimental observations of scatterer transparency and the merging of $\mu_{eff}(E_{eff})$ at large n for short range limited scattering would further validate the modeling of radiation-induced degradation as an independent mechanism utilizing Matthiessen's rule. Nevertheless, the modeling approach presented here can be directly applied to calculations described in [21,22,23] for model radiation-induced degradation in GFET *I-V* characteristics.

5. CONCLUSIONS

In this work, we have investigated TID effects in top-gated epitaxial GFETs. Oxide charge trapping near the SiO_2/graphene channel interface has been identified as the main contributing mechanisms to radiation-induced degradation. The effects of the radiation-induced oxide trapped charges are manifested as negative voltage shifts in the *I-V* characteristics and carrier mobility degradation. These effects have been observed for devices exposed in air as well as for devices exposed with flowing ultra pure N_2 purging moisture contaminants from the irradiation vessel. Extractions of shifts in the Dirac voltage (ΔV_{Dirac}) indicate buildup of oxide trapped charge densities (ΔN_{ot}) of ~ 10^{12} cm^{-2} at dose levels of 1000 krad(Si). The experimental results also reveal a significant increase in minimum conductivity (σ_{min}) for devices tested in air. Since σ_{min} was initially low in these samples (*i.e.*, $\sigma_{min0} = 4.7\ e^2/h$) the increase may be due to compensation and/or reorganization of the potential fluctuations in the graphene layer as a result of molecular adsorption at the SiO_2 surface near the graphene channel [5].

The results reported in this work confirm that GFET devices with top-gated structures are also susceptible to TID effects that result from the buildup of oxide-trapped charges. Mitigation of these degradation mechanisms is required for the successful implementation of GFET devices in space-based applications. An example of such a mitigation technique based on local etching of the dielectric layer in the active regions of back-gated GFETs is described in [5]. However, GFET devices with independent top-gates are desirable for the fabrication of integrated circuits. The devices used in this work were research test structures with relatively thick gate dielectrics. Alternative mitigation approaches such as the using thin top-gate dielectric layers may be available soon as advances in the processing of graphene devices allows further scaling.

An approach for extracting the radiation-induced degradation of the effective carrier mobility from the degraded *I-V* characteristics of irradiated GFETs is presented in this paper. This approach accounts for shifts in V_{Dirac} as well as changes in the minimum conductivity by introducing an intrinsic charge density component (*i.e.*, q_{int}) into the calculations of the total graphene layer charge (*i.e.*, Q_{gc}). The extractions of effective electron mobility (μ_{eff}) indicate deviations from the pre-irradiation $\mu_{eff}(E_{eff})$ dependence that are due to the increased contribution of scattering mechanisms resulting from oxide-trapped charges. For large values of effective vertical field (E_{eff}) the extractions merge at the pre-irradiation curve as mechanisms other than charged impurity scattering become dominant for large carrier densities (*i.e.*, $n \propto E_{eff}$). Similar results were observed recently for Cu-CVD back-gated GFETs irradiated with Co-60 gamma rays [24]. Additionally, an increase in μ_{eff} is observed for $E_{eff} < 1$ MV/cm that is consistent with the observations made in [19] and attributed to scatterer transparency that occurs as carrier wavelengths becomes larger than scatterer spacing. The presented approach can be extended to incorporate interface traps, non-uniform buildup of N_{ot} and the effects of contact resistance as needed based on further experimental investigation.

Additionally, this work introduces a semi-empirical approach for modeling radiation-induced degradation of μ_{eff} based on calculations of E_{eff} and Q_{gc} that incorporate the buildup of N_{ot}. The degradation of μ_{eff} due to charge impurity scattering is modeled as an independent scattering mechanism according to Matthiessen's rule. The appearance of scatterer transparency is modeled empirically as a function of E_{eff} resulting in good qualitative agreement with the experimental extractions of μ_{eff} and its radiation response.

ACKNOWLEDGMENTS

This work was supported (in part) by the Defense Threat Reduction Agency, Basic Research Award # HDTRA1-10-1-0015, to the USC Information Sciences Institute and the University of Southern California.

REFERENCES

1. Geim, K.; Novoselov, K.S. The rise of graphene. *Nat. Mater.* **2007**, *6*, 183–191.
2. Schwierz, F. Graphene transistors. *Nat. Nanotechnol.* **2010**, *5*, 487–496.
3. Meric, I.; Han, M.Y.; Young, A.F.; Ozyilmaz, B.; Kim, P.; Shepard, K.L. Current saturation in zero-bandgap, top-gated graphene field-effect transitors. *Nat. Nanotechnol.* **2008**, *3*, 654–659.
4. Zhang, E.X.; Newaz, A.K.; Wang, B.; Bhandaru, S.; Zhang, C.X.; Fleetwood, D.M.; Bolotin, K.I.; Pantelides, S.T.; Alles, M.L.; Schrimpf, R.D.; *et al.* Low-energy X-ray and ozone-exposure induced defect formation in graphene materials and devices. *IEEE Trans. Nucl. Sci.* **2011**, *58*, 2961–2967.
5. Cress, C.D.; McMorrow, J.J.; Robinson, J.T.; Landi, B.J.; Hubbard, S.M.; Messenger, S.R. Radiation effects in carbon nanoelectronics. *Electronics* **2012**, *1*, 23–31.
6. Mai-Xing, H.; Zhuo-Yu, J.; Li-Wei, S.; Ying-Ping, C.; Hong, W.; Xin, L.; Dong-Mei, L.; Ming, L. Gamma radiation caused graphene defects and increased carrier density. *Chin. Phys. B* **2011**, *20*, 086102.

7. Zhang, C.X.; Zhang, E.X.; Fleetwood, D.M.; Alles, M.L.; Schrimpf, R.D.; Song, E.B.; Sung, M.K.; Galatsis, K.; Wang, K.L.W. Electrical stress and total ionizing dose effects on graphene-based non-volatile memory devices. *IEEE Trans. Nucl. Sci.* **2012**, *59*, 2974–2978.

8. Cress, C.; McMorrow, J.; Robinson, J.; Friedman, A.; Landi, B. Radiation effects in single-walled carbon nanotube thin-film-transistors. *IEEE Trans. Nucl. Sci.* **2010**, *57*, 3040–3045.

9. Fu, Y.; Zhang, J.; Wang, C.; Zhou, C.; Wormuth, R.; Tyree, V.; Fritze, M.; McMarr, P.; Hughes, H. Total Dose Radiation Effects in Carbon Nanotube Transistors. In Presented at the Government Microcircuit Applications and Critical Technology Conference, Las Vegas, NV, USA, March 2012.

10. Moon, J.S.; Curtis, D.; Bui, S.; Hu, M.; Gaskill, D.K.; Tedesco, J.L.; Asbeck, P.; Jernigan, G.G.; VanMil, B.L.; Myers-Ward, R.L.; *et al.* Top-gated epitaxial graphene FETs on Si-Face SiC Wafers With a peak transconductance of 600 mS/mm. *IEEE Electron Dev. Lett.* **2010**, *31*, 260–262.

11. Russo, S.; Craciun, M.F.; Yamamoto, M.; Tarucha, S.; Morpurgo, A.F. Double-gated graphene-based devices. *New J. Phys.* **2009**, *11*, 095018.

12. Chen, J.-H.; Jang, C.; Adam, S.; Fuhrer, M.S.; Williams, E.D.; Ishigami, M. Charge impurity scattering in graphene. *Nat. Phys.* **2008**, *4*, 377–381.

13. Kim, W.; Javey, A.; Vermesh, O.; Wang, Q.; Li, Y.; Dai, H. Hysteresis caused by water molecules in carbon nanotube field-effect transistors. *Nano Lett.* **2003**, *3*, 193–198.

14. Sun, S.C.; Plummer, J.D. Electron mobility in inversion and accumulation layers on thermally oxidized silicon surfaces. *IEEE J. Solid-State Circ.* **1980**, *27*, 1497–1508.

15. Wehling, T.O.; Şaşıoğlu, E.; Friedrich, C.; Lichtenstein, A.I.; Katsnelson, M.I.; Blügel, S. Strength of effective coulomb

interactions in graphene and graphite. *Phys. Rev. Lett.* **2011**, *106*, 236805.

16. Martin, J.; Akerman, N.; Ulbricht, G.; Lohmann, T.; Smet, J.H.; von Klitzing, K.; Yacoby, A. Observation of electron-hole puddles in graphene using a scanning single-electron transistor. *Nat. Phys.* **2008**, *4*, 144–148.
17. Chen, J.-H.; Cullen, W.G.; Jang, C.; Fuhrer, M.S.; Williams, E.D. Defect scattering in graphene. *Phys. Rev. Lett.* **2009**, *102*, 236805.
18. Adam, S.; Hwang, E.H.; Galitski, V.M.; Sarma, S.D. A self-consistent theory for graphene transport. *Proc. Natl. Acad. Sci. USA* **2007**, *104*, 18392–18397.
19. Farmer, D.B.; Perebeinos, V.; Ling, Y.; Dimitrakopoulos, C.; Avouris, P. Charge trapping and scattering in epitaxial graphene. *Phys. Rev. B* **2011**, *84*, 205417.
20. Cheng, B.; Woo, J. Measurement and modeling of the n-channel and p-channel MOSFET's inversion layer mobility at room and low temperature operation. *J. Phys. IV France* **1996**, *6*, 43–47.
21. Thiele, S.A.; Schaefer, J.A.; Schwierz, F. Modeling of graphene metal-oxide-semiconductor field-effect transistors with gapless large-area graphene channels. *J. Appl. Phys.* **2010**, *107*, 094505.
22. Meric, I.; Baklitskaya, N.; Kim, P.; Shepard, K.L. RF Performance of Top-Gated, Zero-Bandgap Graphene Field-Effect Transistors. In Proceedings of Electron Devices Meeting, San Francisco, CA, USA, December 2008; pp. 1–4.
23. Champlain, J.G. A first principles theoretical examination of graphene-based field effect transistors. *J. Appl. Phys.* **2011**, *109*, 084515.

24. Cress, C.D.; Champlain, J.G.; Esqueda, I.S.; Robinson, J.T.; Friedman, A.L.; McMorrow, J. Total ionizing dose induced charge carrier scattering in graphene devices. *IEEE Trans. Nucl. Sci.* **2012**, *59*, 3045–3053.

Chapter 8

Radiation Effects in Carbon Nanoelectronics

Cory D. Cress [1,*], Julian J. McMorrow [2], Jeremy T. Robinson [1], Brian J. Landi [3], Seth M. Hubbard [4] and Scott R. Messenger [1]

[1]*Electronics Science & Technology Division, U.S. Naval Research Laboratory, Washington, DC 20375, USA*

[2]*Global Strategies Group (North America) Inc., Crofton, MD 21114, USA*

[3]*Chemical & Biomedical Engineering Department, Rochester Institute of Technology, Rochester, NY 14623, USA*

[4]*Physics Department, Rochester Institute of Technology, Rochester, NY 14623, USA*

ABSTRACT

We experimentally investigate the effects of Co-60 irradiation on the electrical properties of single-walled carbon nanotube and graphene field-effect transistors. We observe significant differences in the radiation response of devices depending on their irradiation environment, and confirm that, under controlled conditions, standard dielectric hardening approaches are applicable to carbon nanoelectronics devices.

Keywords: single walled carbon nanotubes; graphene; total ionizing dose; TID; radiation hardening; FET; SWCNT; carbon nanoelectronics

1. INTRODUCTION

The unique properties of carbon nanomaterials, in particular single-walled carbon nanotubes (SWCNTs) and graphene, have garnered much attention due to their potential incorporation into high-performance devices. Recent laboratory-demonstrated threshold frequencies of 80 GHz (SWCNT based) and 100 GHz (graphene based) field-effect transistors (FETs) establish their position among extremely high-mobility materials. Their high mobility and unique ambipolar transport behavior may also prove useful in other analog devices such as power amplifiers, high-frequency mixers, and radio receivers [1]. It is clear that carbon nanoelectronics will enable, in the near future, leap-ahead technologies to address the needs of terrestrial and space-bound systems. Therefore, it is imperative that the radiation response and methods for radiation hardening be considered in parallel with device development to ensure reliability in harsh environments.

Many theoretical and experimental studies of SWCNTs, and to a lesser extent graphene, have investigated the effects of ionizing radiation on the crystalline structure of these materials [2,3,4]. The results indicate that ionizing radiation can damage the crystalline lattice by creating defects, although the defect formation probability strongly depends on the energy, mass, and angle of the incident ionizing radiation. Recent studies which monitor the electronic transport properties of carbon-based FETs support these conclusions, that is, a high tolerance to proton irradiation (e.g., SWCNT-FETs [5,6], graphene-FETs [7]) or high-energy photon irradiation (e.g., SWCNT [8,9,10,11], graphene-FETs [12,13,14]), where the transport properties of the carbon nanostructures are maintained. However, the overall device response to irradiation is mixed. Some studies show enhanced drain currents (e.g., [7,10]), or insignificant performance changes [5], while others report dramatic changes in device behavior including large shifts in threshold voltage [9]. These assorted results motivate the need for utilizing standardized test structures and controlling the experimental conditions to allow for a systematic evaluation of the radiation response of the materials and devices independent of extrinsic effects.

In the current study, we illustrate the sensitivities of carbon nanoelectronic devices to total ionizing dose (TID) effects caused by both intrinsic and extrinsic factors. We begin by comparing the effects of TID on a back-gated graphene FET irradiated in vacuum and in air. We then highlight an approach to increasing the performance and radiation tolerance of SWCNT thin-film-transistors (TFTs) using thinned, high-κ gate dielectrics.

2. EXPERIMENTAL SECTION

We employ various device structures to achieve the goals listed above, all of which are processed utilizing the NRL NanoScience Institute cleanroom facilities. The preparation of SWCNT thin films (98% semiconducting-enriched, NanoIntegris, Inc.) and subsequent SWCNT-TFT photolithographic processing closely follow our previously reported procedure found in ref. [8,11]. For our back-gated graphene FETs, we grow and transfer graphene films following methods described by Li *et al.* [15] and use similar photolithographic processing as in Reference [8]. Briefly, graphene is grown using chemical vapor deposition (CVD) on Cu foils at 1000 °C in a horizontal tube furnace in flowing H_2 and CH_4. Following growth, graphene is transferred using the following steps: (i) spin-on a protective poly(methyl methacrylate) (PMMA) film; (ii) etch the Cu foil; (iii) transfer the floating PMMA-coated graphene to a substrate of choice; and (iv) remove the PMMA by submersion in an acetone bath. The use of transferred films in the preparation of devices allows us to change the substrate material and structure independently from the carbon nanomaterials. We perform Raman spectroscopy (laser wavelength of 532 nm) of the graphene channel material pre- and post Co-60 irradiation to assess changes in defect concentration in graphene.

2.1. Back-Gated Device Structures

The standard back-gated structure depicted in Figure 1a is commonly used to study the electrical transport properties of SWCNTs and graphene. The structure allows for rapid fabrication of devices with variable dimensions. Furthermore, any semiconductor or metal substrate with an insulating dielectric layer may serve as the global back-gate electrode to modulate the Fermi level of the carbon active layer. The devices studied here use thermally grown SiO_2 on p^+-Si as the starting substrate.

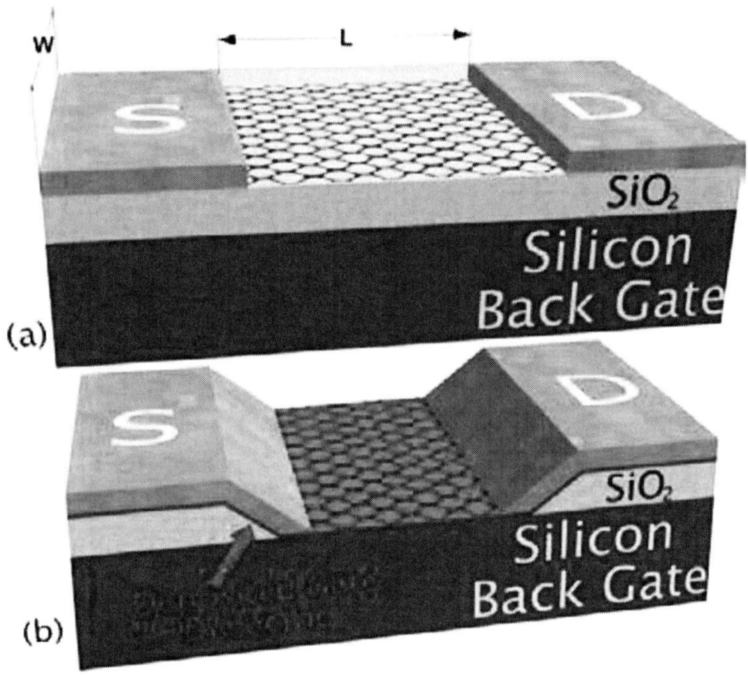

Figure 1. (**a**) Schematic depicting a standard back-gated graphene field-effect transistor (FET) and (**b**) a locally etched back-gated FET structure with a deposited dielectric layer. For the single-walled carbon nanotube (SWCNT)-thin-film-transistors (TFTs), a thin mesh of SWCNTs replaces the graphene film depicted here.

Though versatile, these back-gated structures are not optimized for high performance since they typically have thick dielectric layers (>1000 Å) resulting in low gate capacitance, and thus, require biases of tens-of-volts to fully saturate or pinch-off the drain current. Furthermore, thick dielectric layers are highly susceptible to radiation induced charge build-up, which is known to cause threshold voltage shifts and increased leakage in metal oxide semiconductor (MOS) devices [16]. To mitigate these effects, we locally etch the dielectric layer in the active region of the back-gated FET, which is depicted in Figure 1b. We then deposit a gate dielectric material (e.g., Al_2O_3 or Si_3N_4; depicted in red in Figure 1b) over the entire substrate with a thickness controlled by growth. This design feature is important since high quality dielectric layers deposited directly onto SWCNTs or graphene for use as a top gate electrode are still under development. Using the locally etched back-gated (LEBG) structure allows us to maintain the properties of the transferred carbon nanomaterial channel, and have flexibility in the choice and processing performed on the dielectric layer prior to transferring the channel material and fabricating devices.

2.2. SiO_2 Local Etching

We pattern the channel regions of the LEBG devices using standard photolithographic procedures. These substrates are etched for 1–3 min in buffered oxide etch (BOE) (7:1 NH_4F:HF), followed by a short ~1 min rinse in flowing deionized water. This etchant produces smooth hydrogen-terminated Si terraces [17,18], thereby providing a controlled surface for subsequent dielectric layer deposition.

2.3. Si_3N_4 Plasma Enhanced Chemical Vapor Deposition (PECVD)

We deposit Si_3N_4 films on LEBG substrates and BOE etched control wafers for optical and electrical characterization using an Oxford Instruments PlasmaPro™ System 100. Freshly prepared substrates are loaded, pumped to 5×10^{-6} Torr and preheated to 350 °C for 10 min.

Si_3N_4 deposition proceeds at a pressure of 2 Torr with an N_2 carrier gas, and SiH_4 and NH_4 as the Si and N precursors, respectively. We obtain an index of refraction of 1.97 at 633 nm based on fitted reflectance measurements indicating nearly stoichiometric Si_3N_4. Following SWCNT thin-film deposition, annealing the structures in air at 300 °C for 16 h introduces oxygen into the nitride layer transforming it into silicon oxynitride (SiON) with an index of refraction of 1.75 at 633 [11]. On control samples, we deposit Ti/Au contact pads to perform capacitance and electrical breakdown measurements. For a 23 nm SiON film we obtained an average dielectric constant of ε_r = 5.5 at 1 MHz and dc-breakdown fields in excess of 8 MV/cm [11].

2.4. Al_2O_3 Atomic Layer Deposition (ALD)

We deposit Al_2O_3 films on LEBG substrates and BOE etched control wafers for optical and electrical characterization using an Oxford Instruments FlexAL® system. In a similar fashion as the Si_3N_4 we deposit Al_2O_3on freshly prepared substrates, which are loaded, pumped, and preheated to 300 °C. The growth proceeds at this temperature as trimethyl aluminum (TMA) the Al precursor and oxygen plasma are alternatingly pulsed into the growth chamber. We deposited Ti/Au contact pads to perform capacitance and electrical breakdown measurements. For a 32 nm Al_2O_3 film we obtained an average dielectric constant of ε_r = 8.3 at 1 MHz and dc-breakdown fields in excess of 10 MV/cm.

3. RESULTS AND DISCUSSION

3.1. Characterizing Air Sensitivity of Carbon Electronics

We recently observed that the response of SWCNT-TFTs to Co-60 exposure is highly dependent on the local environment of the device during irradiation [8]. We report here a similar experiment on a standard back-gated graphene FET with a 250 nm SiO_2 gate oxide,

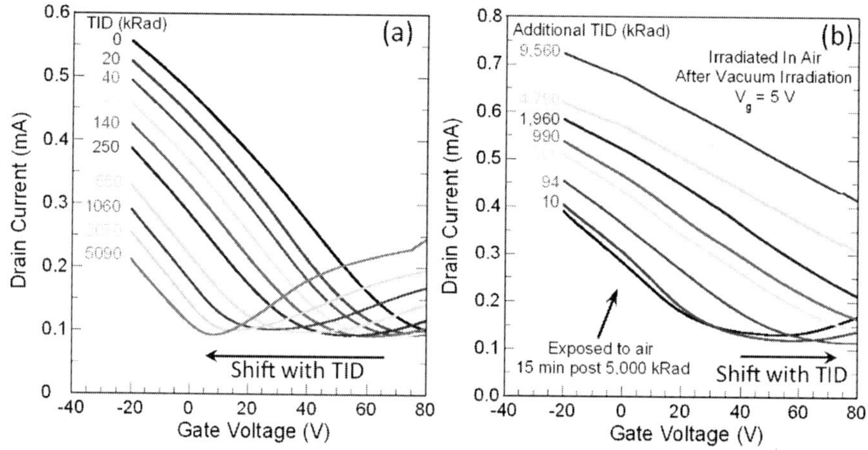

Figure 2. (**a**) Transfer characteristics of a back-gated graphene-FET with incremental total ionizing dose; (**b**) Transfer characteristics of the same device following 15 min of air exposure (black curve) and with additional total ionizing dose (TID).

channel length of 15 μm, channel width of 60 μm, and a positive gate bias of 5 V during irradiation. Figure 2a illustrates the TID dependent transfer characteristics of the wire bonded standard back-gated graphene FET measured in static vacuum. We measured the pre-irradiation transfer characteristics (black trace labeled 0 krad(Si)) immediately before the first TID exposure (1 krad = 10 Gy). We performed control measurements which yielded constant transfer characteristics for the device held in the sample chamber under static vacuum prior to irradiation. We observe a large shift in the transfer characteristics (towards negative V_g) following a TID of 20 krad(Si), which proceeds with further dosing. Following a TID of 40 krad(Si), the Dirac point becomes apparent at $V_g = 73$ V along with the onset of the electron transport for $V_g > 73$ V. These results are consistent with hole trapping in the SiO_2, which increases the Fermi level closer to mid-gap. Following the initial Co-60 irradiation up to 5 Mrad(Si), we exposed the device to air for 15 m then repeated the experiment with the same device in an air ambient as shown in Figure 2b. During the 15 m repose, the transfer characteristics shift towards positive gate voltage, resulting from room temperature annealing of the radiation-induced trapped charges and due to adsorption of molecular species from the air. Additional irradiation in an air ambient causes the transfer characteristics to shift further towards positive gate bias, the opposite of what we observed when irradiated under vacuum. This indicates that molecular species from the air, namely oxygen and water, overpower the effects of SiO_2 trapped holes and decrease the Fermi level promoting hole doping in the channel [14].

Using the standard long-channel field effect mobility equation: $\mu_{FE} = \frac{L/W}{C_{ox}V_{DS}}\frac{dI_D}{dV_G}$, where L is the channel length, W is the width, C_{ox} is the oxide capacitance, I_D is the drain current, and V_G is the gate voltage, we calculate the peak μ_{FE} from the transfer curves as plotted verses TID in Figure 3a for the graphene FET irradiated in vacuum and air. When irradiated in vacuum, the hole μ_{FE} initially degrades from 1090 cm^2/Vs to 980 cm^2/Vs following a TID of 200 krad(Si), recovers slightly then degrades again reaching a minimum mobility of 896 cm^2/Vs after a TID of 5 Mrad(Si). During the 15 m repose, the mobility recovers considerably from the low of 896 cm^2/Vs increasing to 1126 cm^2/Vs exceeding the pre-irradiation mobility. With increasing TID exposure in air, the mobility decreases to 907 cm^2/Vs after a TID of 100 krad(Si), then after a brief plateau, degrades to 663 cm^2/Vs following a total (additional) TID of 5 Mrad(Si) in air.

We also plot the variation in σ_{min} with TID for the transfer curves measured in vacuum in Figure 3a (lower right *y*-axis). In this figure, σ_{min} displays non-monotonic behavior that approximately follows that of the mobility (measured in vacuum). The σ_{min}, for devices of these dimensions, provides a relative measure of the charge inhomogeneity at the graphene-substrate interface resulting from trapped charges in SiO_2 [19] and adsorbed impurities including oxygen, moisture, and photoresist residues [20]. The initial decrease in $\sigma_{min}$with TID reflects the increasing trapped charge density and magnitude of charge potential fluctuations (*i.e.*, electron-hole puddles) within the graphene channel. These fluctuations restrict current flow in the graphene channel, which favors transport through regions of unperturbed potential, and is expected to reduce the mobility as we observe. Following a TID of 200–500 krad(Si), the σ_{min} and μ_{FE} begin to recover. We attribute this behavior to a reorganization of the potential fluctuations in the graphene, potentially resulting in a correlated charge distribution in the SiO_2 [21] or rearrangement of mobile surface adsorbates, though additional work is needed to confirm this mechanism.

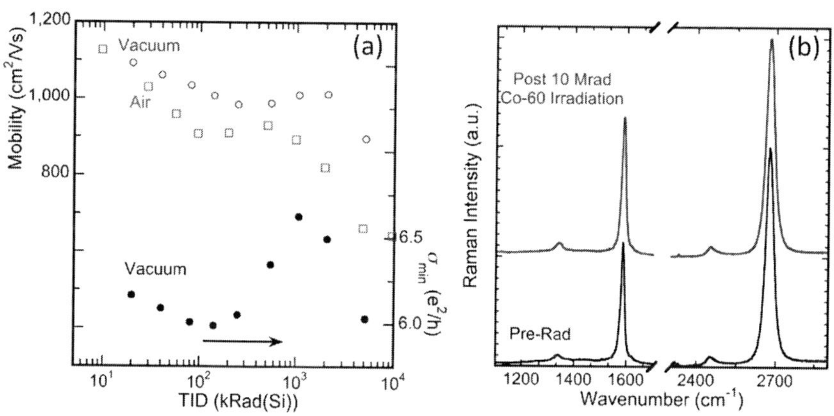

Figure 3. (**a**) Field effect mobility as a function of TID for a graphene-FET irradiated in vacuum (blue, open circles) and in air (red, open squares), along with the minimum conductivity (*i.e.*, the conductivity at the Dirac point) for the transfer characteristics measured in vacuum; (**b**) Representative Raman spectra of the graphene FET both pre- and post 10 Mrad(Si) irradiation.

In Figure 3b we provide characteristic pre- and post 10 Mrad(Si) Co-60 irradiation Raman spectra of the graphene device measured in Figure 2. We measure a 2D-band (2670 cm^{-1}) to G-band (1580 cm^{-1}) peak intensity ratio of 1.5, which is characteristic of single layer graphene on SiO_2 [22]. The D-band mode (~1345 cm^{-1}) results from disorder in the graphene, potentially arising from defects, edges, wrinkles, or residual PMMA resist, but the relatively low intensity indicates the disorder is minimal. Furthermore, we observe no significant difference in the D-band intensity following irradiation indicating that the changes observed in the graphene FETs due to Co-60 irradiation result from trapped charges in the SiO_2 gate oxide and not due to changes in lattice defect concentration. These results are consistent with our recent study on SWCNTs-TFTs and emphasize the need to control the environmental conditions of carbon electronics, especially when investigating the basic radiation response mechanisms [8]. Furthermore, the controlled environment results (vacuum data) confirm that, like Si MOS-FETs, carbon electronics are susceptible to the oxide trapped charge effects necessitating the development of radiation-hardened gate dielectrics.

3.2. SiON and Al_2O_3 Gate Dielectrics with the Locally Etched Back-Gated Structure (SWCNT-TFTs)

In Figure 4a we compare the transfer characteristics of two LEBG SWCNT-TFTs with either a 32 nm SiON or a 32 nm Al_2O_3 gate dielectric layer, 2 µm channel lengths, and 32 µm channel widths. Both devices are representative of the devices with drain current $I_{d,on}/I_{d,off}$ greater than 10^4 (about 20 devices for each gate dielectric). For both dielectric layers, we see a maximum drain current of 1×10^{-7} A and a consistently lower minimum drain current for the SiON (about 10× lower). The larger off-current may result from the onset of electron transport in the Al_2O_3 device reflecting the ambipolar transport properties of SWCNTs [23]. A different level of adsorbed water due to the different surface energies of Al_2O_3 and Si_3N_4 may cause this, which is supported by the fact that similar ambipolar behavior is observed in LEBG SWCNT-TFTs with a SiON gate dielectric following vacuum annealing at 125 °C and 1×10^{-5} Torr (see Figure 4b).

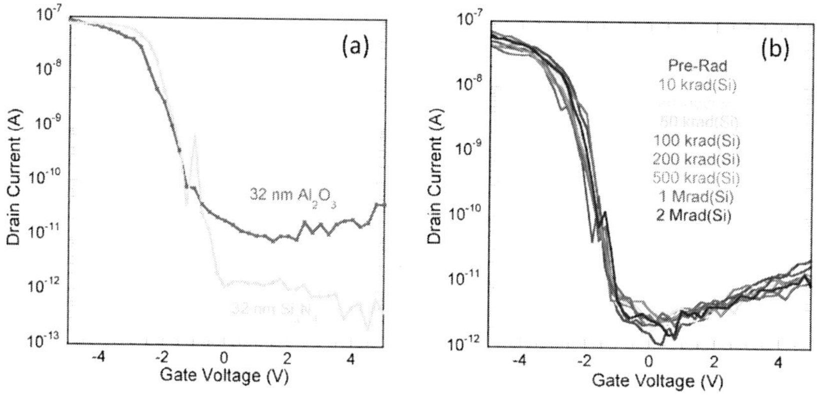

Figure 4. (**a**) Transfer characteristics of SWCNT-TFTs with 32 nm Al_2O_3 (red) and with 32 nm SiON (green) gate dielectric layers; (**b**) The radiation response of SWCNT-TFT with a 23 nm SiON gate dielectric layer.

In Figure 4b we investigate the effects of Co-60 irradiation for a LEBG SWCNT-FET with a 23 nm SiON gate oxide and channel length and width of 2 μm and 16 μm, respectively, biased with a 0.25 MV/cm gate field during irradiation. By scaling the gate dielectric material to 23 nm we observe essentially no effect of Co-60 irradiation on the transfer characteristics up to a TID of 2 Mrad(Si). We observe a maximum shift in the threshold voltage of −0.25 V at a TID of 100 krad(Si) after which the transfer characteristics begin to shift back towards positive $V_g$values. The stability in V_g results from lower trapped charge accumulation in the thin SiON layer—the thinner layer allows more carriers to escape by tunneling or through field assisted transport [24]. Furthermore, the trapping characteristics of SiON are distinct from SiO_2 and favor electron trapping over hole trapping [16]. Therefore, we attribute the positive shift in threshold voltage to electron traps in the SiON [11], although it is not possible to completely rule out some molecular doping during irradiation in static vacuum conditions.

The data we have presented here is promising evidence towards hardening carbon nanoelectronics against TID exposure. Therefore, our future investigations are aimed at maintaining the radiation tolerance levels while simultaneously improving device performance and reducing the susceptibility to atmospheric doping effects.

4. CONCLUSIONS

We have demonstrated the sensitivities of graphene FETs, including shifting Dirac point and mobility degradation, due to TID exposure as well as the compounded effects caused by doping from molecular adsorbates. We observe a correlation between the minimum conductivity, σ_{min}, and hole mobility, μ_{FE}, both of which vary non-monotonically with increasing TID. We attribute this behavior to an evolving electron-hole puddle distribution, controlled by SiO_2 trapped charges and mobile surface impurities. Using a locally etched back-gate region, we have created a device structure that can leverage standard hardening approaches including the use of alternative dielectric materials and thinner gate dielectric layers. This has led to the demonstration of a SWCNT-TFT which is nearly unaffected by ionizing radiation up to a total dose of 2 Mrad(Si).

ACKNOWLEDGMENTS

We acknowledge financial support of this research from the Defense Threat Reduction Agency. We would like to thank H. Hughes, P. McMarr, D. McMorrow for helpful discussions, and A. Friedman for wire bonding.

REFERENCES

1. Kocabas, C.; Kim, H.-S.; Banks, T.; Rogers, J.A.; Pesetski, A.A.; Baumgardner, J.E.; Krishnaswamy, S.V.; Zhang, H. Radio frequency analog electronics based on carbon nanotube transistors. *Proc. Natl. Acad. Sci. USA* **2008**, *105*, 1405–1409.

2. Cress, C.D.; Schauerman, C.M.; Landi, B.J.; Messenger, S.R.; Raffaelle, R.P.; Walters, R.J. Radiation effects in single-walled carbon nanotube papers. *J. Appl. Phys.* **2010**, *107*, 014316.

3. Krasheninnikov, A.V.; Nordlund, K. Ion and electron irradiation-induced effects in nanostructured materials. *J. Appl. Phys.* **2010**, *107*, 071301.

4. Kaiser, A.B.; Skákalová, V.; Roth, S. Modelling conduction in carbon nanotube networks with different thickness, chemical treatment and irradiation. *Phys. E* **2008**, *40*, 2311–2318.

5. Hong, W.-K.; Lee, C.; Nepal, D.; Geckeler, K.; Shin, K.; Lee, T. Radiation hardness of the electrical properties of carbon nanotube network field effect transistors under high-energy proton irradiation.*Nanotechnology* **2006**, *17*, 5675–5680.

6. Comfort, E.S.; Fishman, M.; Malapanis, A.; Hughes, H.; McMarr, P.; Cress, C.D.; Bakhru, H.; Lee, J.U. Creation of Individual Defects at Extremely High Proton Fluences in Carbon Nanotube Diodes. *Nucl. Sci.***2011**, *58*, 2898–2903.

7. Ko, G.; Kim, H.Y.; Ren, F.; Pearton, S.J.; Kim, J. Electrical characterization of 5 MeV proton-irradiated few layer graphene. *Electrochem. Solid State Lett.* **2010**, *13*, K32–K34.

8. Cress, C.D.; McMorrow, J.J.; Robinson, J.T.; Friedman, A.L.; Landi, B.J. Radiation effects in single-walled carbon nanotube thin-film-transistors. *IEEE Trans. Nucl. Sci.* **2010**, *57*, 3040–3045

9. Tang, X.; Yang, Y.; Kim, W.; Wang, Q.; Qi, P.; Dai, H. Measurement of ionizing radiation using carbon nanotube field effect transistor. *Phys. Med. Biol.* **2005**, *50*, N23–N31.

10. Vitusevich, S.A.; Sydoruk, V.A.; Petrychuk, M.V.; Danilchenko, B.A.; Klein, N.; Offenhäusser, A.; Ural, A.; Bosman, G. Transport properties of single-walled carbon nanotube transistors after gamma radiation treatment. *J. Appl. Phys.* **2010**, *107*, 063701.

11. Cress, C.D.; McMorrow, J.J.; Robinson, J.T.; Friedman, A.L.; Hughes, H.L.; Weaver, B.D.; Landi, B.J. Total ionizing dose-hardened carbon nanotube thin-film transistors with silicon oxynitride gate dielectrics. *MRS Commun.* **2011**, *1*, 27–31.

12. Cress, C.D.; McMorrow, J.J.; Robinson, J.T.; Landi, B.J.; Hubbard, S.M.; Messenger, S.R. Radiation-hardening of carbon nanoelectronics. In Proceedings of the Government Microcircuit Applications and Critical Technologies Conference, Orlando, FL, USA, 2011; pp. 1–5.

13. Schrimpf, R.D.; Fleetwood, D.M.; Alles, M.L.; Reed, R.A.; Lucovsky, G.; Pantelides, S.T. Radiation effects in new materials for nano-devices. *Microelectron. Eng.* **2011**, *88*, 1–6.

14. Zhang, E.X.; Newaz, A.K.M.; Wang, B.; Bhandaru, S.; Zhang, C.X.; Fleetwood, D.M.; Bolotin, K.I.; Pantelides, S.T.; Alles, M.L.; Schrimpf, R.D.; *et al.* Low-energy X-ray and ozone-exposure induced defect formation in graphene materials and devices. *IEEE Trans. Nucl. Sci.* **2011**, *58*, 2961–2967.

15. Li, X.; Zhu, Y.; Cai, W.; Borysiak, M.; Han, B.; Chen, D.; Piner, R.; Colomba, L.; Ruoff, R. Transfer of large-area graphene films for high-performance transparent conductive electrodes. *Nano Lett.* **2009**, *9*, 4359–4363.

16. Oldham, T. *Ionizing Radiation Effects in MOS Oxides*; World Scientific: Singapore, 1999.

17. Le Thanh, V.; Bouchier, D.; Débarre, D. Fabrication of SiGe quantum dots on a Si(100) surface. *Phys. Rev. B***1997**, *56*, 10505–10510.

18. Pietsch, G.J. Hydrogen on Si: Ubiquitous surface termination after wet-chemical processing. *Appl. Phys. A***1995**, *60*, 347–363.

19. Sui, Y.; Low, T.; Lundstrom, M.; Appenzeller, J. Signatures of disorder in the minimum conductivity of graphene. *Nano Lett.* **2011**, *11*, 1319–1322.

20. Adam, S.; Hwang, E.H.; Galitski, V.M.; Das Sarma, S. A self-consistent theory for graphene transport. *Proc. Natl. Acad. Sci. USA* **2007**, *104*, 18392–18397.

21. Yan, J.; Fuhrer, M. Correlated charged impurity scattering in graphene. *Phys. Rev. Lett.* **2011**, *107*, 206601.

22. Li, X.; Cai, W.; An, J.; Kim, S.; Nah, J.; Yang, D.; Piner, R.; Velamakanni, A.; Jung, I.; Tutuc, E.; *et al*. Large-area synthesis of high-quality and uniform graphene films on copper foils. *Science* **2009**, *324*, 1312–1314.

23. Javey, A.; Shim, M.; Dai, H. Electrical properties and devices of large-diameter single-walled carbon nanotubes. *Appl. Phys. Lett.* **2002**, *80*, 1064.

24. Saks, N.; Ancona, M.; Modolo, J. Radiation Effects in MOS Capacitors with Very Thin Oxides at 80 °K. *IEEE Trans. Nucl. Sci.* **1984**, *31*, 1249–1255.

Chapter 9

Carbon Nano Material Synthesis from Polyethylene by Chemical Vapour Deposition

Pravin Jagdale[1], Madhuri Sharon[2], Golap Kalita[3], Noor Mahmad Nabi Maldar[1], Maheshwar Sharon[2*]

[1]*Chemistry Department, Solapur University, Solapur, India*

[2]*N. Shankaran Nair Research Centre for Nanotechnology and Bionanotechnology, S.I.C.E.S. College, Ambarnath, India*

[3]*Department of Electrical Engineering, Chubu University, Kasugai, Japan*

ABSTRACT

Three different types of Polyethylene family, High Density Polyethylene, (HDPE), Low Density polyethylene (LDPE) and Linear Low Density polyethylene (LLDPE) polymers having different molecular weight and density; were pyrolyzed in the temperature range of 550°C - 1050°C under H_2, N_2 and Ar gases. Taguchi Optimization technique was applied to find out the best operating conditions to get maximum yield of carbon nano material (CNM). For Taguchi optimization, experimental set up was done in two different temperature ranges i.e. 550°C - 750°C and 850°C - 1050°C. CNMs synthesized were characterized by SEM, TEM, Micro Raman and XRD analysis. HDPE was found to yield maximum CNM. Its pyrolysis at 750°C under hydrogen atmosphere for 2h gave carbon nano beads and some carbon nano tubes. Whereas under same conditions at 1050˚C more multi wall carbon nano tubes (MWCNT) were produced, with some carbon nano beads. XRD data confirmed the graphitic nature of carbon-nanotube. The intensities of G-band and D-band of Raman spectra suggested that CNM has more defect sites and spectra were similar for CNM obtained in both the temperature ranges. The TGA analysis of CNM obtained at 550˚C - 750˚C, indicated that they are not amorphous carbon and CNM obtained at 850˚C - 1050˚C decomposed at 624˚C - 668˚C; suggesting that CNT synthesized at this temperature range were more crystalline than what was obtained at the 550˚C - 750˚C.

Keywords: Polyethylene; Pyrolysis; LDPE; HDPE; LLDPE; Carbon Nano Material (CNM); Carbon Nano Beads (CNB); Carbon Nano Tubes (CNT)

1. INTRODUCTION

Discovery of carbon nano material (CNM) in 1991 [1] has attracted enormous attention of scientists; because of their unique structural, chemical, mechanical and electrical properties. In order to utilize CNM for commercial applications, efforts are being made to reduce the cost of its production. Because of its simplicity, CVD technique has been found to be the most popular technique for the large scale production of CNM. One of the factors controlling the price of the CNM is the precursors used for its production. The conventionally used precursors are expensive petro-derivatives e.g. benzene, acetylene etc. However, plant derived waste products like camphor, oils of non-edible plants [2-5] have also been tried, in our lab as source of cheaper precursor.

In our pursuit of getting a cheaper precursor we have directed our efforts to use waste plastics; especially polythenes of thickness of less than 50 μm that are creating environmental disposal problems. Since these materials are composed of polyethylene (PE) and contains long linear or branched chains of hydrocarbons (which are a good precursor for synthesis of CNM) it was thought to be appropriate to develop the pyrolysis technology to synthesize CNM from waste plastics. This attempt will not only solve the problem of disposal of waste plastic in most ecofriendly way (as it converts plastic in absence of oxygen, thus does not produce oxidized toxic carbon compounds like carbon monoxide and dioxane) but also converts the waste (plastic) to wealth (CNT). There has been efforts made in this direction, but due to proprietor and patentable nature of the work, there are very few publications available.

In order to establish conditions of synthesis of CNM from waste plastics, it was decided to develop the technique for the synthesis of CNM first from virgin PE polymers. This was necessary because the waste plastic may contain mixture of different types of polymers. In this paper results obtained with only virgin polymers are discussed; though we have successfully synthesized CNT using same methodology from waste PE also. Application for getting patent has been applied; hence detail cannot be disclosed at present.

Since there were more than 3 parameters involved for pyrolysis techniques and each parameter was with 3 variables, it was decided to use Taguchi Optimization Technique to find out the most suitable parameters of pyrolysis that can give maximum yield of CNM [6-8].

1.1. Taguchi Optimization Methodology

In a pyrolysis process the product's quality and quantity varies widely due to the parameters like temperature, flow rate and type of carrier gas, catalyst and precursor etc. Each parameters can have several levels. For example, temperature can have a range from 400˚C to 900˚C. Likewise other parameters will also have various levels. Considering these factors, the number of experiments becomes enormous to select the most effective parameters which can give the desired product. Taguchi Optimization Technique gives an opprotunity to carry out minimum number of experiments yet giving the best desired parameters and its level to get the desired product [9,10].

Taguchi method is based on calculating the statistical analysis of "Orthogonal Arrays" (OA) of data [11,12]. The ANOVA is used to analyse the results of the OA experiment to determine how much variation in each parameters has contributed towards the production of desired product. This is found out by calculating the Signal-to-Noise ratio. Signal-to-Noise ratios (S/N) are log functions of desired output. If our interest is in getting product with largest quantity, then "larger the better" is used for such calculation. This is calculted by using the Equation (1):

$$\mathrm{S/N} = -10\log_{10}\left[(1/n)\sum 1/y_i^2\right] \tag{1}$$

where "y_i" is the mean response calculated as $y = 1/n\sum y_i$ and n is the number of experiments carried out under similar conditions.

1.2. Calculation for Effect of Each Parameter

To determine the effect of each parameter level (m_i) average value of S/N ratios is calculated for each parameter using "Analysis of Mean" (ANOM) [9-13]. For this calculation the S/N ratios of each experiment with corresponding parameter levels are calculated using Equation (2).

$$m_i = (1/n)\sum \mathrm{S/N} \tag{2}$$

Two types of average value of S/N ratio are calculated. One is overall mean S/N ratio calculated from the entire experiments viz. from nine experiments of L9 orthogonal array (Tables 1 and 2). The second average S/N ratio is calculated for each parameter from Equation (1). Sum of Squares (SOS) are calculated by using Equation (3),

$$\mathrm{SOS} = \sum n\left(m_i - \langle m_i \rangle\right)^2 \tag{3}$$

where, $\langle m_i \rangle$ is average value of m_i.

The total Degrees of Freedom (DOF) is considered as (number of levels of paprameters minus one). If there are three levels for each parameters then DOF = 2.

Factor of Effect (FOE) are calculated from Equation (4),

$$FOE = \frac{SOS}{DOF \times (SOS/DOF)} \tag{4}$$

From this FOE the percentage effect of each parameter on the production of the desired product is calculated.The main advantage of this technique is that, if for example there are 4 parameters each having 3 variables, and if they are to be repeated for 3 times, the total number of experiments becomes very large. Taguchi Optimization requires only 27 experiments if they are to be repeated three times. It is for these advantages, we have used Taguchi Optimization technique in the present work.

2. MATERIAL AND METHODS

Pyrolysis of three types of PE i.e. High Density Polyethylene, (HDPE), Low Density polyethylene (LDPE) and Linear Low Density polyethylene (LLDPE) polymers was performed. The parameters considered were 1) two different ranges of temperature i.e. lower temperature range (550°C - 750°C) and higher temperature range (850°C - 1050°C); 2) the carrier gases namely, hydrogen, nitrogen and argon and 3) duration of pyrolysis namely 1, 2 and 3 h. In order to calculate the impact of parameters on the production of CNM, factor effect was also calculated. Details of the methodology for the calculations of Taguchi Optimization technique are discussed elsewhere [6-8].

Different Parameters & Their Levels for L-9 Orthogonal Matrix

As per Taguchi optimization methodology, two tables were constructed from the various combinations of all possible parameters with a minimum of three variables of each parameter. Variable parameters were selected which are orthogonal to each other. These parameters were: pyrolysis temperature, duration, carrier gas and type of polyethylene. Since, Taguchi Optimization technique can use only three levels of a parameter; two separate orthogonal tables were constructed, one orthogonal table utilizing 550˚C - 750˚C and the other table using 850˚C - 1050˚C temperature. Thus, two L9 orthogonal tables were constructed by using the parameters and their levels as mentioned in Tables 1 and 2. Nano powder of iron was used as Catalyst and it was kept same for all experiments.

Table 1. The orthogonal array of parameters to be optimized by Taguchi method and their impact on yield of CNM at 550˚C - 750˚C (Table also shows the percentage yield of carbon nanomaterial obtained and calculated values of signal to noise ratio).

Expt. No.	Pyrolysis Conditions and Yield of CNM					
	Polymer	Temperature (˚C)	Gas	Duration (min)	CNM	
					% yield	S/N
L1	LDPE	550	H_2	60	0.9	–20.9
L2	LDPE	650	N_2	120	0.2	–33.9
L3	LDPE	750	Ar	180	3.6	–8.8
L4	HDPE	550	N_2	180	0.1	–40
L5	HDPE	650	Ar	60	14.0	2.9
L6	HDPE	750	H_2	120	28.5	9.1
L7	LLDPE	550	Ar	120	0.1	–40
L8	LLDPE	650	H_2	180	5.2	–5.7
L9	LLDPE	750	N_2	60	3.5	–9.1

Table 2. The orthogonal array of parameters to be optimized by Taguchi method and their impact on yield of CNM at 850°C - 1050°C (Table also shows the percentage yield of carbon nanomaterial obtained and calculated values signal to noise ratio).

Expt. No.	Pyrolysis Conditions and Yield of CNM					
	Polymer	Temperature (°C)	Gas	Duration (min)	CNM	
					% yield	S/N
L1	LDPE	850	H_2	60	12.8	2.14
L2	LDPE	950	N_2	120	6	–4.4
L3	LDPE	1050	Ar	180	0.3	–30
L4	HDPE	850	N_2	180	1.5	–16
L5	HDPE	950	Ar	60	17.0	4.61
L6	HDPE	1050	H_2	120	19.3	5.71
L7	LLDPE	850	Ar	120	11.5	1.21
L8	LLDPE	950	H_2	180	0.8	–22
L9	LLDPE	1050	N_2	60	5.6	–5

Pyrolysis was carried out for each set of condition as mentioned in Tables 1 and 2. Percentage yield of CNM and its S/N ratio were calculated for each experimental result by using Equation 1. The yields of CNM as well as S/N ratio calculated from each set are also given in Tables 1 and 2. CNM was analyzed by SEM, TEM, Micro Raman Spectroscopy, XRD and TGA.

3. RESULTS AND DISCUSSIONS

3.1. Taguchi Optimized Pyrolysis Results of Temperature Range 550°C - 750°C

Signal/Noise (S/N) ratio of CNM produced under different conditions is plotted in **Figure 1**. Since our interest was to find out the condition of pyrolysis which could give the maximum amount of CNM, "Larger the Better" condition was used to calculate the S/N ratio. All those levels of parameters for which S/N ratio was more positive to the mean value of S/N ratio were considered as suitable levels for the production of CNM (**Figure 1**(a)). Amongst three types of PE polymers, HDPE appeared to give the best yield. The calculation of % impact of each parameter on the production of CNM (**Figure 1**(b)) suggests that maximum impact was that of the temperature (51.69%) followed by gas (25.12%) and then the duration (15.14%). It was interesting to note that when the pyrolysis was done in the temperature range of 850˚C - 1050˚C duration of pyrolysis had highest impact (84.8%) on production of carbon nano tubes (**Figure 2**(b)).

SEM analysis—as it can be seen in **Figure 2**(a), at 750˚C, pyrolysis under hydrogen atmosphere for 2 h duration yielded a mixture of carbon nano beads with some amount of carbon nano tubes.

TEM micrograph—of CNT confirmed that the CNTs are multi wall carbon nano tubes (**Figure 3**).

XRD peaks at 26.35 and 43.56 of carbon materials, obtained from pyrolyzing HDPE at 750˚C (**Figure 4**(a)) suggests CNM to possess graphitic nature along with some amorphous carbon.

Raman spectra (**Figure 4**(b)) of all the samples showed almost similar results, with D-band around 1335 cm^{-1} and G-band around 1511.11 cm^{-1}. The intensity of Gband is less than that of D-band suggesting that CNM has more defect sites. The position of peak for G-band is more towards the D-band also supports this conclusion.

3.2. Taguchi Optimized Pyrolysis Results of Temperature Range 850˚C - 1050˚C

S/N ratio—calculation of CNM obtained by the pyrolysis of HDPE in the temperature range of 850˚C - 1050˚C shows that maximum yield of CNM was obtained at 1050˚C in presence of hydrogen atmosphere for pyrolysis duration of 2 h (**Figure 5**(a)). It is interesting to note that while for 550˚C - 750˚C; temperature showed the highest % impact on the production of CNM, whereas in the range of 850˚C - 1050˚C, temperature had least impact i.e. only 3.5% and duration of pyrolysis had more impact of 84.78% (**Figure 5**(b)). Similar results have been observed earlier also while using plant material as precursors [8].

XRD—peaks (**Figure 6**(a)) of CNM obtained from pyrolyzing HDPE at 850˚C, confirms the CNM to be graphitic carbon (26.3˚ 2θ).

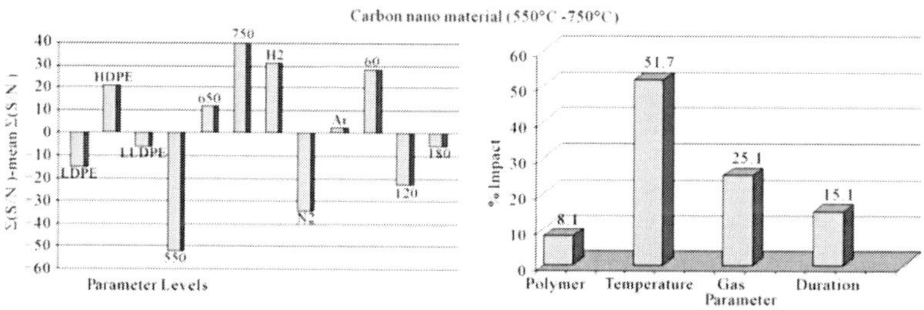

(a) (b)

Figure 1. (a) Graph showing the variation in signal to noise ratio for each level of corresponding parameters. Mean value of signal to noise ratio has been taken as zero and calculation is done with "larger the better"; (b) Shows the % impact of each parameters.

(a)

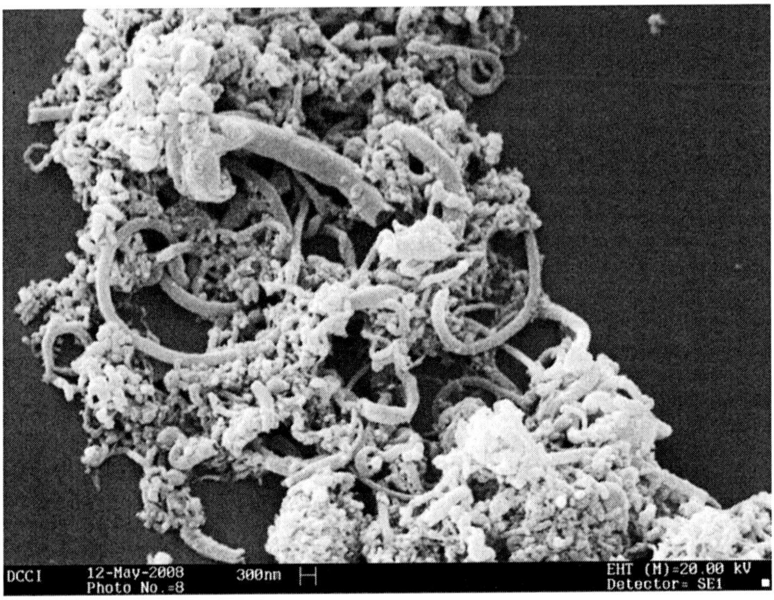

(b)

Figure 2. SEM micrographs of CNM obtained by pyrolyzing HDPE in hydrogen atmosphere for 2 h at (a) 750˚C showing mixture of carbon nano beads and some carbon nano tubes and (b) 1050˚C showing more carbon nano tubes with some carbon nano beads.

Raman spectra were almost same for CNM produced in the temperature range of 850˚C - 1050˚C. Therefore, a typical Raman spectra obtained from CNT at 850˚C is presented in **Figure 6**(b), which shows D-band at 1332 cm^{-1} indicating defect in graphitic carbon. Further Gband at 1587 cm^{-1} is the characteristic peak for graphitic carbon. It is interesting to note that D-band obtained with carbon produced either at 750˚C or 850˚C are almost the same (~1332 - 1335 cm^{-1}), which appears at the expected peak of carbon containing defects (1332 cm^{-1}). However, the G-band peaks obtained with carbon produced at 750˚C is at 1511 cm^{-1} while for carbon produced at 850˚C is 1587 cm^{-1}. This suggests that while the defect present in both forms of these carbons are almost the same type but graphitization has taken place more at higher temperature.

SEM micrographs of CNM obtained at 550˚C - 750˚C, and those obtained in the temperature range of 850˚C - 1050˚C are shown in Tables 3 and 4 along with their TGA spectra. It interesting to observe that CNM obtained at lower temperature range i.e. 550˚C - 750˚C, most of them are carbon beads with very small amount of carbon nanotubes. But at higher temperature beads seems to have elongated to CNTs among with some remnant carbon nano beads, hence there is a large percentage of carbon nanotubes and very few carbon nano beads. Taguchi optimization methodology has also suggested that yield of carbon nano tubes is affected more by the temperature (**Figure 1**(b)) or duration of exposure to temperature (**Figure 5**(b)) during of pyrolysis; as also noted earlier with other precursors [8]. The SEM micrographs of Figure 2 suggest that carbon nano beads and carbon nano tubes are obtained when HDPE is pyrolyzed at 750˚C and 1050˚C respectively.

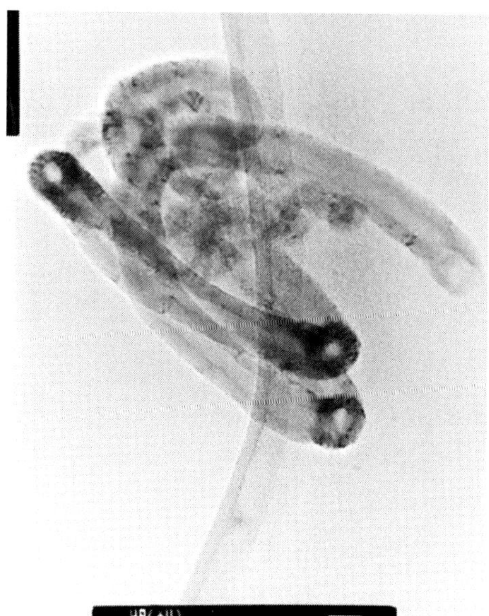

Figure 3. TEM micrographs of CNT (obtained by pyrolyzing HDPE in hydrogen atmosphere for 2 h at 1050˚C showing coiled multi walled carbon nano tubes.

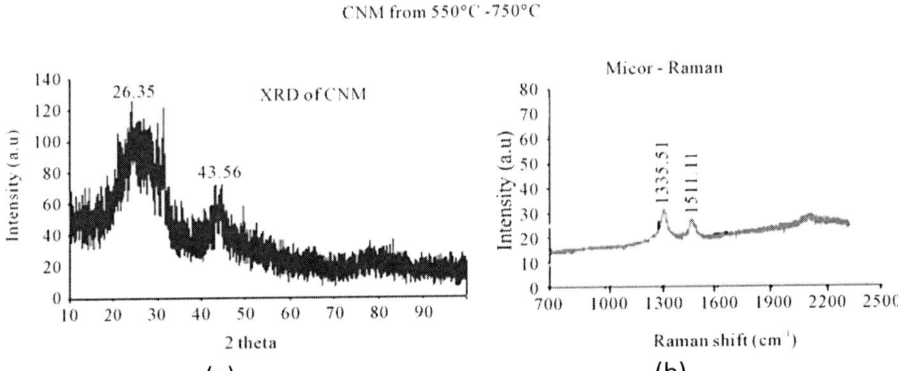

(a) (b)

Figure 4. (a) XRD of the CNM produced by HDPE at 750˚C shows peak at 26.35 (002) and 43.56 which are characteristic peak of the graphitic material; (b) Micro Raman spectra of CNM obtained in the temperature range 550˚C -750˚C.

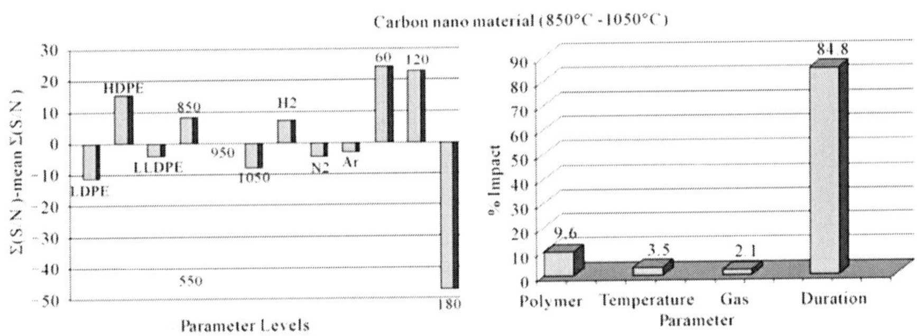

(a) (b)

Figure 5. (a) Graph showing the variation of signal to noise ratio for each level of corresponding parameters. Mean value of signal to noise ratio has been taken as zero and calculation is done with "larger the better" (850˚C - 1050˚C); (b) Showing the % impact of each parameters.

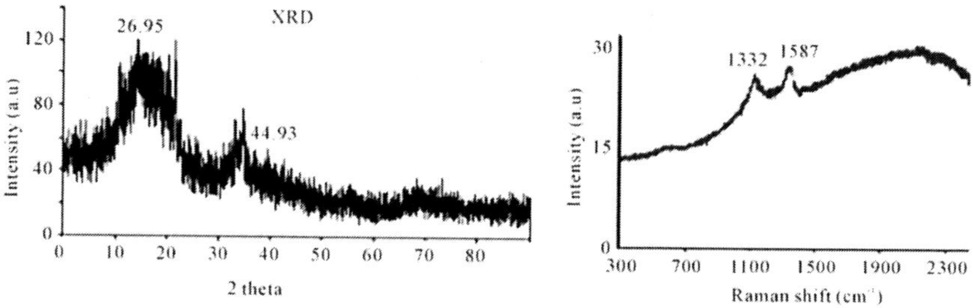

(a) (b)

Figure 6. (a) XRD analysis of the un-purified CNM from LDPE at 850°C showing characteristic peaks of the graphitic material; (b) Micro Raman of CNM obtained from 850°C - 1050°C showing D-band at 1332 cm^{-1}, indicating defect in graphitic carbon.

	SEM	TGA
L1	NA	NA
L2		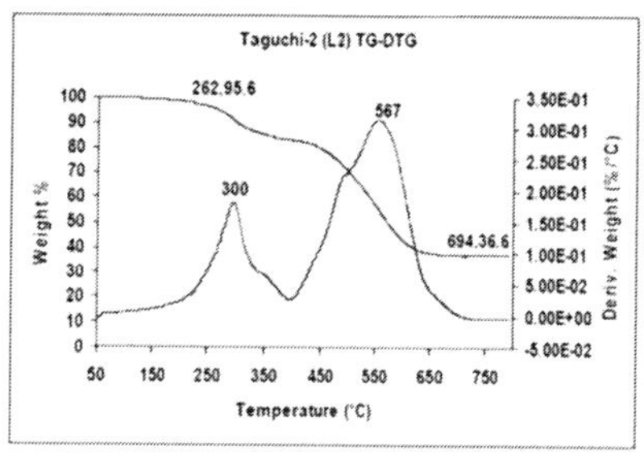
L3		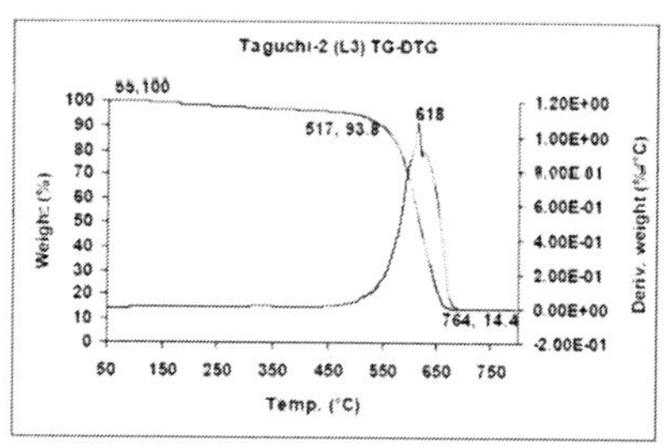
L4	NA	NA

L5

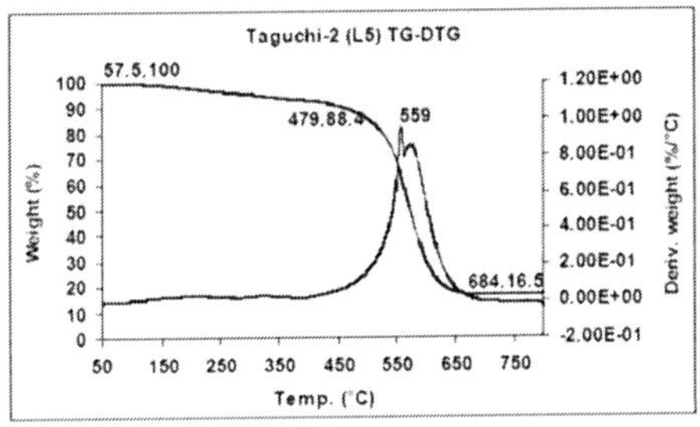

L6

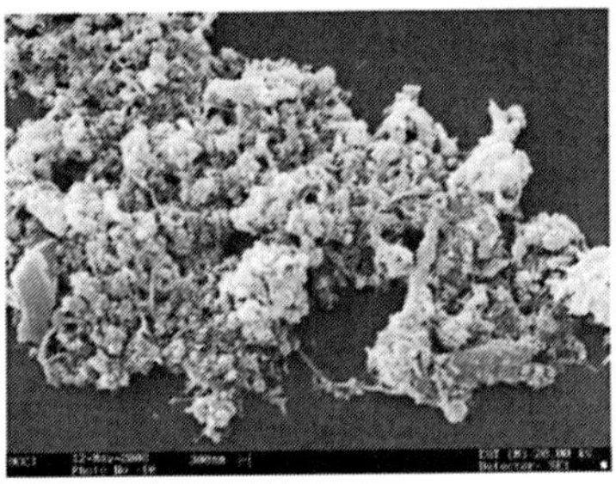

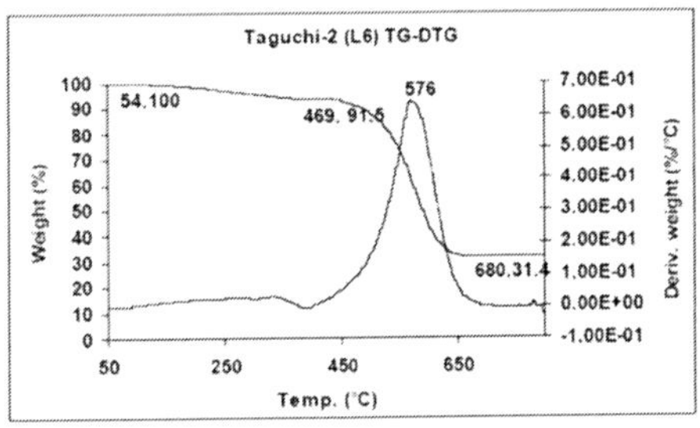

L7 NA NA

L8

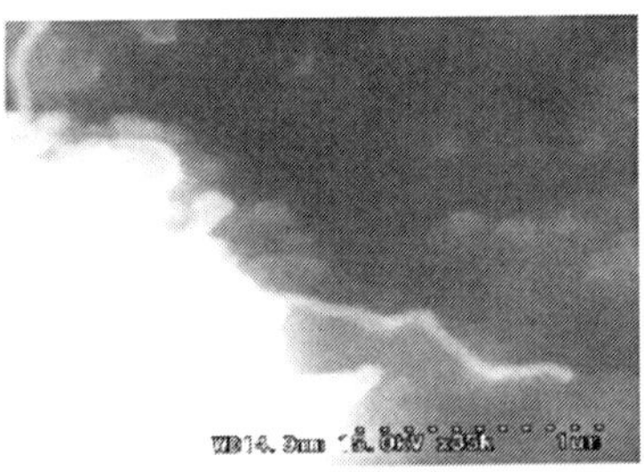

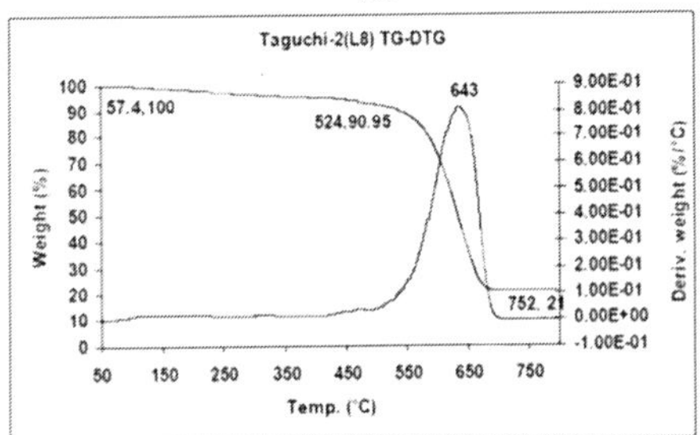

L9

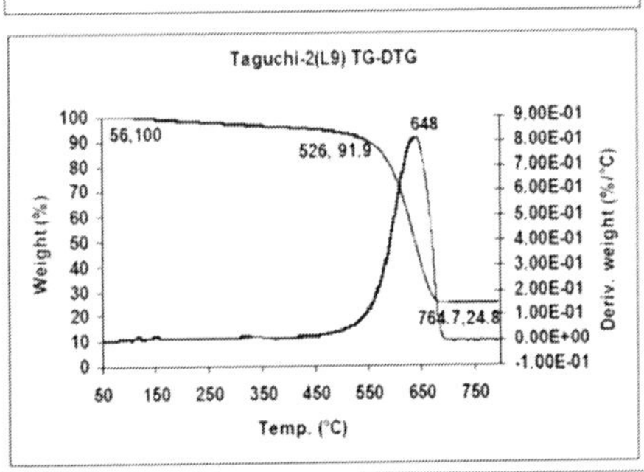

Table 3. CNM Taguchi method (550˚C - 750˚C) analysis by SEM and TGA.

	SEM	TGA

L1

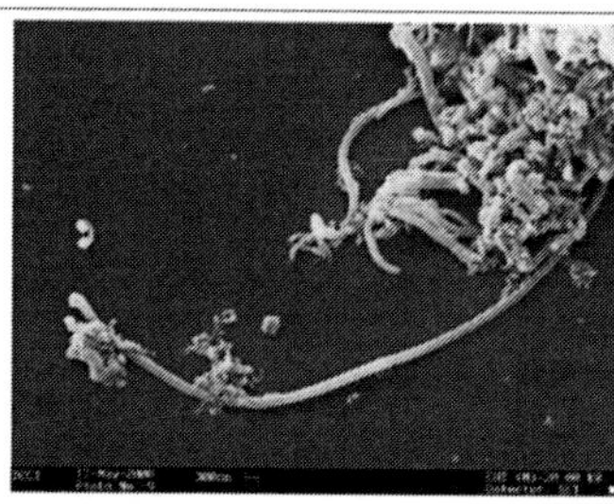

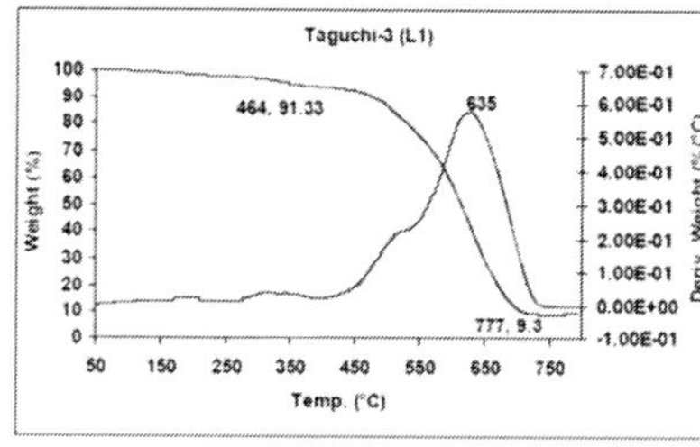

L2

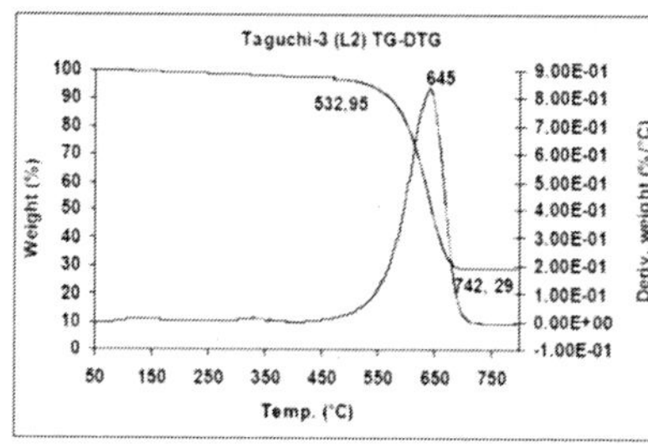

L3

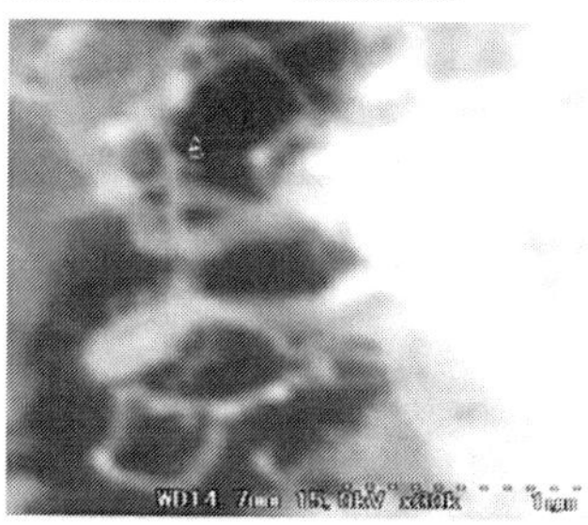

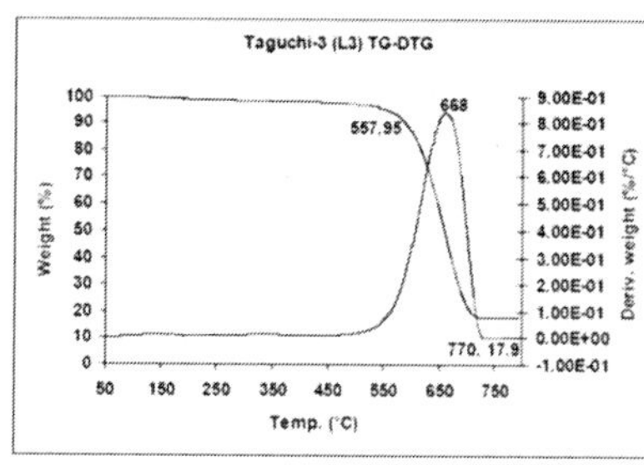

L4

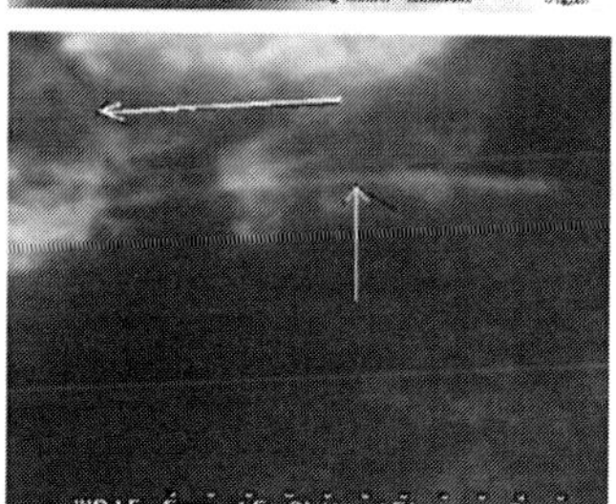

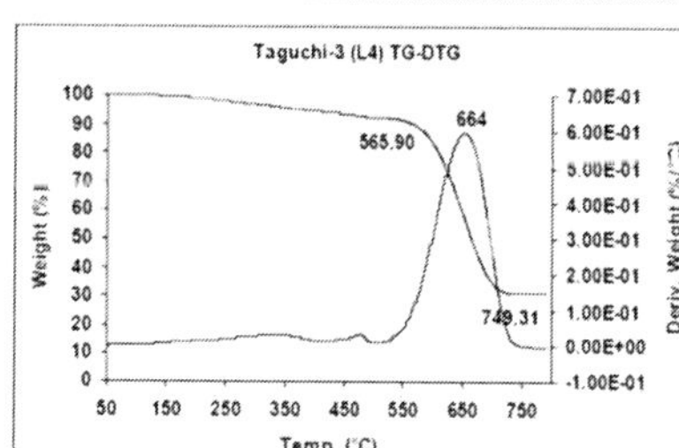

L5

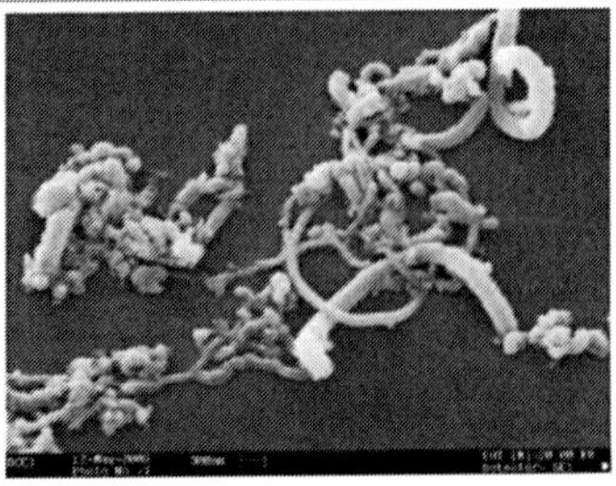

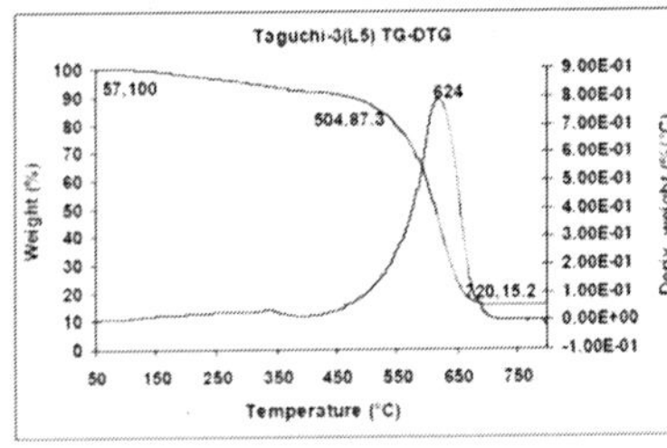

L6

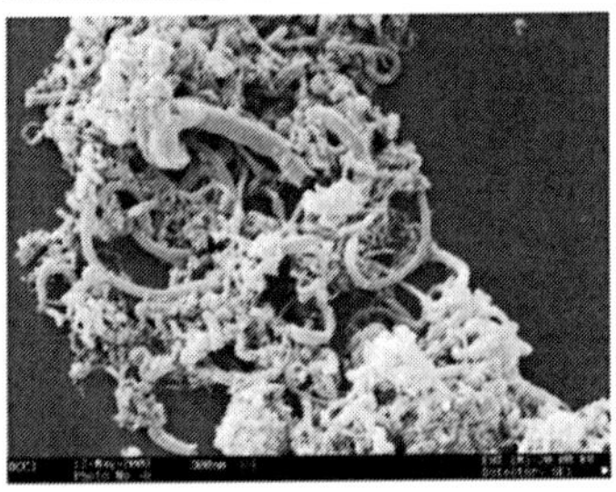

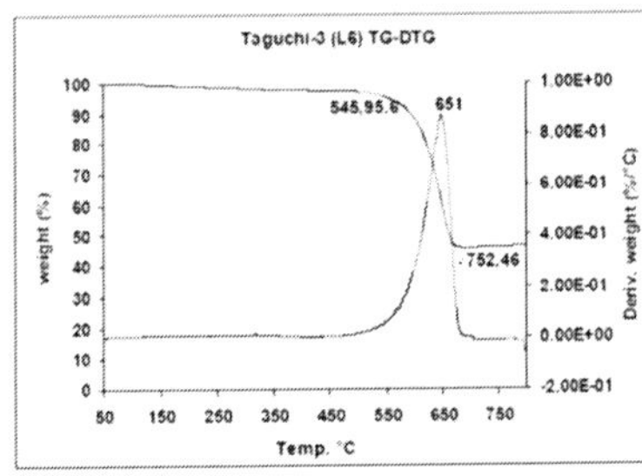

L7

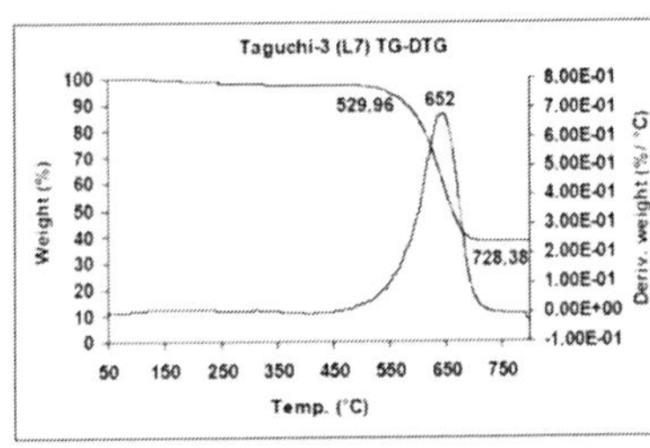

L8

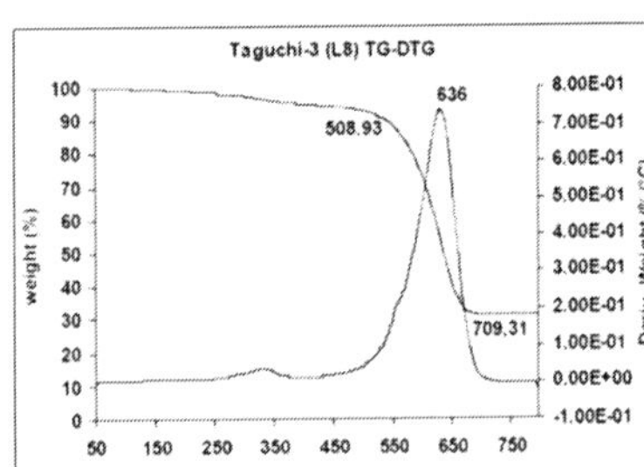

L9

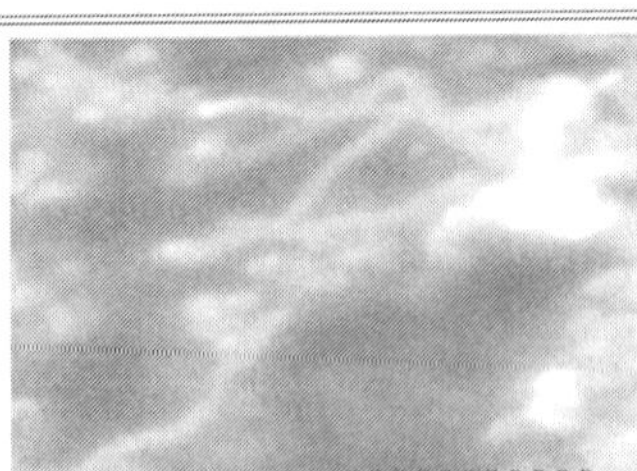

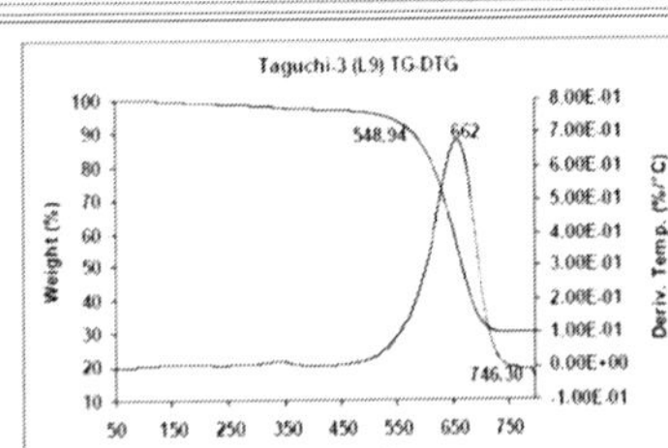

Table 4. CNM Taguchi method (850°C - 1050°C) analysis by SEM and TGA.

TGA Results of the derivatives of TGA also showed that CNM obtained at 550˚C - 750˚C, decomposed in the temperature range of 559˚C - 648˚C which is an indication that these materials are not amorphous carbon (Table 3). Derivatives of TGA of CNM obtained at 850˚C - 1050˚C shows that these materials decomposes at 624˚C - 668˚C; suggesting that CNT synthesized at this temperature range is more crystalline than what was obtained at the 550˚C - 750˚C (Tables 3 and 4). Raman spectra also support this observation.

4. CONCLUSION

Virgin polymers (LDPE, HDPE and LLDPE) on pyrolysis using CVD system can be converted into MWCNT or CNB depending on the temperature range used. Taguchi Optimization methodology helped in selecting the best suitable parameters for high yield of desired CNM. Hydrogen atmosphere and a duration of 2 h pyrolysis was found to be the best condition for both low (550˚C - 750˚C) temperature range that produced CNB; as well as high (850˚C - 1050˚C) temperature range, which produced more MWCNT. Various characterization data viz. SEM, TEM, Micro RAMAN & TGA; all supported the same findings that CNM synthesized were graphitic, crystalline material having MWCNT or CNB morphology. Temperature higher than 600˚C helped in getting rid of amorphous carbon from the CNM produced.

5. ACKNOWLEDGEMENTS

Authors acknowledge UGC, New Delhi for financial support to carryout this project.

REFERENCES

1. S. Iijima, "Helical Microtubules of Graphitic Carbon," Nature, Vol. 354, 1991, pp. 56-58.
2. M. Sharon, K. Mukhopadhyay and K. M. Krishna, "Fullerenes from Camphor: A Natural Source," Physical Review Letters, Vol. 72, No. 20, 1994, pp. 3182-3185.
3. M. Sharon, D. Pradhan, M. Kumar and Y. Ando, "NanoOctopus: A New Form of Branching Carbon Nanofiber," Journal of Nanoscience and nanotechnology, Vol, 3, No. 3, 2003, pp. 1-3.
4. M. Sharon, "Cabon Nanomaterials," Encyliopedia Nanoscience and Nanotechnology, Vol. 1, 2004, pp. 517-546.
5. A. Afre. Rakesh, T. Soga, T. Jimbo, Mukul Kumar, Y. Ando and M. Sharon, "Growth of Vertically Aligned Carbon Nanotubes on Silicon and quartz Substrate by Spray Pyrolysis of a Natural Precursor; Turpentine Oil," Chemical Physics Letters, Vol. 414, No. 1-3, 2005, pp. 6-10.
6. M. Sharon, P. R. Apte, S. C. Purandara and R. Zacharia, "Application of the Taguchi Analytical Method for Optimization of Effective Parameters of the Chemical Vapour Deposition Controlling the Production of Nanotube/ Nanobeads," Journal of Nanoscience and Nanotechnology, Vol. 5, No. 2, 2005, pp. 288-295.
7. N. Bejoy, M. Sharon and D. K. Mishra, "Application of Taguchi Methodology for Optimization of Parameters of CVD Influencing Formation of a Desired Optical Band Gap of Carbon Film," Carbon Science, Vol. 6, No. 2, 2005, pp. 96-100.
8. S. Bhardwaj S, M. Sharon and T. Ishihara, "Taguchi Optimization of the Carbon Anode for Li-ion Battery from Natural Precursors," Current Applied Physics, Vol. 8, No. 1, 2008, pp. 71-77.

9. S. Maghsoodloo, G. Ozdemir, V. Jorden and C. H. Husang, "Strengths and Limitations of Taguchi's Contributions to Quality, Manufacturing, and Process Engineering," Journal of Manufacturing System, Vol. 23, No. 2, 2004, pp. 73-126.

10. R. H. Lochner, "Designing for Quality: An Introduction to the Best of Taguchi and Western Methods of Statistical Method Design," Chapman and Hall, London, 1990.

11. H. Gitlow, A. Alan Oppenheim and R. Oppenheim, "Tools and Methods for the Improvement of Quality," Von Hoffman Press Inc., Boston, 1989.

12. S. R. Ravella, C. Ganesh, P. R. Shetty and P. J. Hobbs, "The Taguchi Methodology as a Statistical Tool for Biotechnological Applications: A Critical Appraisal," Biotechnology Journal, Vol. 3, No. 4, 2008, pp. 510-523.

13. E. Box, "Statistics for Experiments," Wiley & Sons, New York, 1978.

Chapter 10

Exact Traveling Wave Solutions for Nano-Solitons of Ionic Waves Propagation along Microtubules in Living Cells and Nano-Ionic Currents of MTs

Emad H. M. Zahran

Department of Mathematical and Physical Engineering, College of Engineering, University of Benha, Shubra, Egypt

ABSTRACT

In this work, the extended Jacobian elliptic function expansion method is used as the first time to evaluate the exact traveling wave solutions of nonlinear evolution equations. The validity and reliability of the method are tested by its applications to nano-solitons of ionic waves propagation along microtubules in living cells and nano-ionic currents of MTs which play an important role in biology.

Keywords: Extended Jacobian Elliptic Function Expansion Method, Nano-Solitons of Ionic Waves Propagation along Microtubules in Living Cells, Nano-Ionic Currents of MTs, Traveling Wave Solutions

1. INTRODUCTION

The nonlinear partial differential equations of mathematical physics are major subjects in physical science [1] . Exact solutions for these equations play an important role in many phenomena in physics such as fluid mechanics, hydrodynamics, optics, and plasma physics. Recently many new approaches for finding these solutions have been proposed, for example, tanh-sech method [2] - [4] , extended tanh-method [5] - [7] , sine-cosine method [8] - [10] , homogeneous balance method [11] [12] , F-expansion method [13] - [15] , exp-function method [16] [17] , trigonometric function series method [18] , (G'/G) expansion method [19] - [22] , Jacobi elliptic function method [23] - [26] and so on.

The objective of this article is to apply the extended Jacobian elliptic function expansion method for finding the exact traveling wave solution of nano-solitons of ionic waves propagate on along microtubules in living cells and nano-ionic currents of MTs which play an important role in biology and mathematical physics.

The rest of this paper is organized as follows: In Section 2, we give the description of the extended Jacobi elliptic function expansion method. In Section 3, we use this method to find the exact solutions of the nonlinear evolution equations pointed out above. In Section 4, conclusions are given.

2. DESCRIPTION OF METHOD

Consider the following nonlinear evolution equation

$$F\left(u, u_t, u_x, u_{tt}, u_{xx}, \cdots\right) = 0, \tag{2.1}$$

where F is polynomial in u(x, t) and its partial derivatives in which the highest order derivatives and nonlinear terms are involved. In the following, we give the main steps of this method [23] - [26].

Step 1. Using the transformation

$$u = u(\xi), \xi = x - ct, \tag{2.2}$$

where k and c are the wave number and wave speed, to reduce Equation (2.1) to the following ODE:

$$P(u, u', u'', u''', \cdots) = 0, \tag{2.3}$$

where P is a polynomial in $u(\xi)$ and its total derivatives, while $u' = \mathrm{d}u/\mathrm{d}\xi$.

Step 2. Making good use of ten Jacobian elliptic functions, we assume that (2.3) have the solutions in these forms:

$$u(\dot{\varepsilon}) = a_0 + \sum_{j=1}^{N} f_i^{j-1}(\dot{\varepsilon})\left[a_j f_i(\dot{\varepsilon}) + b_j g_i(\dot{\varepsilon})\right],\ i = 1,2,3,\cdots \tag{2.4}$$

with

$$f_1(\xi) = sn\xi,\ g_1(\xi) = cn\xi,$$

$$f_2(\xi) = sn\xi,\ g_2(\xi) = dn\xi,$$

$$f_3(\xi) = ns\xi,\ g_3(\xi) = cs\xi,$$

$$f_4(\xi) = ns\xi,\ g_4(\xi) = ds\xi,$$

$$f_5(\xi) = sc\xi,\ g_5(\xi) = nc\xi,$$

$$f_6(\xi) = sd\xi,\ g_6(\xi) = nd\xi, \tag{2.5}$$

where $sn\xi$, $cn\xi$, $dn\xi$, are the Jacobian elliptic sine function. The jacobian elliptic cosinefunction and the Jacobian elliptic function of the third kind and other Jacobian functions which is denoted by Glaisher's symbols and are generated by these three kinds of functions, namely

$$ns\xi = \frac{1}{sn\xi},\ nc\xi = \frac{1}{cn\xi},\ nd\xi = \frac{1}{dn\xi},\ sc\xi = \frac{cn\xi}{sn\xi},$$
$$cs\xi = \frac{sn\xi}{cn\xi},\ ds\xi = \frac{dn\xi}{sn\xi},\ sd\xi = \frac{sn\xi}{dn\xi} \quad (2.6)$$

that has the relations

$$sn^2\xi + cn^2\xi = 1,\ dn^2\xi + m^2 sn^2\xi = 1,\ ns^2\xi = 1 + cs^2\xi,$$
$$ns^2\xi = m^2 + ds^2\xi,\ sc^2\xi + 1 = nc^2\xi,\ m^2 sd^2 + 1 = nd^2\xi \quad (2.7)$$

with the modulus m (0 < m < 1): In addition we know that

$$\frac{\mathrm{d}}{\mathrm{d}\xi} sn\xi = cn\xi dn\xi,\ \frac{\mathrm{d}}{\mathrm{d}\xi} cn\xi = -sn\xi dn\xi,\ \frac{\mathrm{d}}{\mathrm{d}\xi} dn\xi = -m^2 sn\xi cn\xi. \quad (2.8)$$

The derivatives of other Jacobian elliptic functions are obtained by using Equation (2.8). To balance the highest order linear term with nonlinear term we define the degree of u as D[u] = n which gives rise to the degrees of other expressions as

$$D\left[\frac{\mathrm{d}^q u}{\mathrm{d}\xi^q}\right] = n + q,\ D\left[u^p\left(\frac{\mathrm{d}^q u}{\mathrm{d}\xi^q}\right)^s\right] = np + s(n+q). \quad (2.9)$$

According the rules, we can balance the highest order linear term and nonlinear term in Equation (2.3) so that n in Equation (2.4) can be determined.

In addition we see that when m → 1, $sn\xi$, $cn\xi$, $dn\xi$ degenerate as $\tanh\xi$, $\mathrm{sech}\xi$, $\mathrm{sech}\xi$, respectively, while when therefore Equation (2.5) degenerate as the following forms

$$u(\xi) = a_0 + \sum_{j=1}^{N} \tanh_i^{j-1}(\xi)\left[a_j \tanh(\xi) + b_j \mathrm{sech}(\xi)\right], \quad (2.10)$$

$$u(\xi) = a_0 + \sum_{j=1}^{N} \coth_i^{j-1}(\xi)\left[a_j \coth(\xi) + b_j \coth(\xi)\right], \quad (2.11)$$

$$u(\xi) = a_0 + \sum_{j=1}^{N} \tan_i^{j-1}(\xi)\left[a_j \tan(\xi) + b_j \sec(\xi)\right], \tag{2.12}$$

$$u(\xi) = a_0 + \sum_{j=1}^{N} \cot_i^{j-1}(\xi)\left[a_j \cot(\xi) + b_j \csc(\xi)\right]. \tag{2.13}$$

Therefore the extended Jacobian elliptic function expansion method is more general than sine-cosine method, the tan-function method and Jacobian elliptic function expansion method.

3. APPLICATION

3.1. Example 1: Nano-Solitons of Ionic Waves Propagation along Microtubules in Living Cells [27]

We first consider an inviscid, incompressible and non-rotating flow of fluid of constant depth (h). We take the direction of flow as x-axis and z-axis positively upward the free surface ingravitational field. The free surface elevation above the undisturbed depth h is η(x; t), so that the wave surface at height z = h + η(x; t), while z = 0 is horizontal rigid bottom.

Let j(x; z; t) be the scalar velocity potential of the fluidlying between the bottom (z = 0) and free space η(x; t), then we could write the Laplace and Euler equation with the boundary conditions at the surface and the bottom, respectively, as follows:

$$\frac{\partial^2 \varphi}{\partial x^2} + \frac{\partial^2 \varphi}{\partial z^2} = 0; \;\; 0 < z < h + \eta; \;\; -\infty < x < +\infty \tag{3.1}$$

$$\frac{\partial \varphi}{\partial t} + \frac{1}{2}\left(\frac{\partial \varphi}{\partial z}\boldsymbol{i} + \frac{\partial \varphi}{\partial z}\boldsymbol{k}\right)^2 + g\eta = 0, \quad z = h + \eta \tag{3.2}$$

$$\frac{\partial \eta}{\partial t} + \frac{\partial \eta}{\partial x}\frac{\partial \varphi}{\partial x} - \frac{\partial \varphi}{\partial z} = 0, \tag{3.3}$$

$$\frac{\partial\varphi}{\partial z}=0;\ \ z=0 \tag{3.4}$$

It is useful to introduce two following fundamental dimensionless parameters:

$$\sigma=\frac{\eta_0}{h}<1;\ \ \delta=\left(\frac{h}{l}\right)^2<1, \tag{3.5}$$

where η_0 is the wave amplitude, and l is the characteristic length-like wavelength. Accordingly, we also take a complete set of new suitable non-dimensional variables:

$$x=\frac{x}{l};\ \ z=\frac{z}{h};\ \ \tau=\frac{ct}{l};\ \ \psi=\frac{\eta}{\eta_0};\ \ \varnothing=\frac{h}{\eta_0 lc}\varphi, \tag{3.6}$$

where   is the shallow-water wave speed, with g being gravitational acceleration.

In term of (3.5) and (3.6) the initial system of Equation (3.1)-(3.4) now reads

$$\delta\frac{\partial^2\varphi}{\partial x^2}+\frac{\partial^2\varphi}{\partial z^2}=0; \tag{3.7}$$

$$\frac{\partial\varnothing}{\partial\tau}+\frac{\sigma}{2}\left(\frac{\partial\varnothing}{\partial x}\right)^2+\frac{\sigma}{2}\left(\frac{\partial\varnothing}{\partial z}\right)^2+\psi=0;\ \ Z=1+\sigma\psi, \tag{3.8}$$

$$\frac{\partial\psi}{\partial\tau}+\sigma\left(\frac{\partial\varnothing}{\partial x}\frac{\partial\psi}{\partial x}\right)-\frac{1}{\delta}\frac{\partial\varnothing}{\partial z}=0;\ \ Z=1+\sigma\psi, \tag{3.9}$$

$$\frac{\partial\varphi}{\partial z}=0;\ \ z=0 \tag{3.10}$$

Expanding $\varnothing(x;t)$ in terms of δ

$$\varnothing = \varnothing_0 + \delta\varnothing_1 + \delta^2\varnothing_2, \tag{3.11}$$

and using the dimensionless wave particles velocity in x-direction, by definition $u = \frac{\partial\varnothing}{\partial x}$, then substituting of

(3.11) into (3.7)-(3.9), with retaining terms up to linear order of small parameters (σ,δ) in (3.8), and second order in (3.9), we get

$$\frac{\partial\varphi_0}{\partial\tau} - \frac{\delta}{2}\frac{\partial^2 u}{\partial\tau\partial x} + \psi + \frac{1}{2}\sigma u^2 = 0, \tag{3.12}$$

$$\frac{\partial\psi}{\partial\tau} + \sigma u\frac{\partial\psi}{\partial x} + \frac{1}{\delta}(1+\sigma\psi)\frac{\partial u}{\partial x} = \frac{\delta}{6}\frac{\partial^3 u}{\partial x^3}. \tag{3.13}$$

Making the differentiation of (3.12) with respect to x, and rearranging (3.13), we get

$$\frac{\partial u}{\partial\tau} + \sigma u\frac{\partial u}{\partial x} + \frac{\partial\psi}{\partial x} - \frac{1}{2}\delta\frac{\partial^3 u}{\partial x^2\partial\tau} = 0, \tag{3.14}$$

$$\frac{\partial\psi}{\partial\tau} + \frac{\partial}{\partial x}\left[u(1+\sigma\psi)\right] - \frac{1}{6}\delta\frac{\partial^3 u}{\partial x^3} = 0. \tag{3.15}$$

Returning back to dimensional variables η(x; t) and $= d\varphi/dx$, (3.14) now reads

$$\frac{\partial v}{\partial t} + v\frac{\partial v}{\partial x} + g\frac{\partial\eta}{\partial x} = \frac{1}{3}h^2\frac{\partial^3 u}{\partial x^2\partial t}. \tag{3.16}$$

We could define the new function V(x, t) unifying the velocity and displacement of water particles as follows:

$$v = \frac{1}{h}\frac{\partial V}{\partial t}; \ \eta = -\frac{\partial V}{\partial x} \tag{3.17}$$

implying that (3.16) becomes

$$\frac{\partial^2 V}{\partial t^2} - gh\frac{\partial^2 V}{\partial x^2} + \frac{1}{2h}\frac{\partial}{\partial x}\left(\frac{\partial V}{\partial t}\right)^2 = \frac{1}{3}h^2\frac{\partial^4 V}{\partial x^2 \partial t^2}. \tag{3.18}$$

We seek for traveling wave solutions with moving coordinate of the form $\xi = x - vt$ and with wave speed v, which reduces Equation (3.18) into ordinary nonlinear differential equation as follows:

$$\left(v^2 - gh\right)\frac{\partial^2 V}{\partial \zeta^2} + \frac{v^2}{2h}\frac{\partial}{\partial \zeta}\left(\frac{\partial V}{\partial \zeta}\right)^2 = \frac{1}{3}h^2 v^2 \frac{\partial^4 V}{\partial \zeta^4} \tag{3.19}$$

Integrating Equation (3.19) once, and setting $\frac{\partial V}{\partial \xi} = W$, we get

$$\frac{\partial^2 w}{\partial \xi^2} = \alpha w^2 + \beta w + c_1 \tag{3.20}$$

Balancing W'' and W^2 yields, N + 2 = 2N → N = 2. Therefore, we can write the solution of Equation (3.20) in the form

$$W(\xi) = a_0 + a_1 sn + b_1 cn + a_2 sn^2 + b_2 sncn \tag{3.21}$$

$$W'(\xi) = \left(a_1 cn - b_1 sn + 2a_2 sncn + b_2 - 2b_2 sn^2\right) dn \tag{3.22}$$

$$\begin{aligned} W''(\xi) = {} & 6a_2 m^2 + sn^4 + 2a_1 m^2 sn^3 - 4a_2\left(m^2+1\right) - a_1\left(m^2+1\right) sn \\ & + 6b_2 sm^2 sn^3 cn + 2b_1 m^2 sn^2 cn - b_2\left(4+m^2\right) sncn - b_1 cn + 2a_2 \end{aligned} \tag{3.23}$$

Substituting (3.21) into (3.23), setting the coefficients of (sn^4, sn^3, sn^3cn, sn^2, sn^2cn, sncn, sn, cn, sn^0) to zero, we obtain the following underdetermined system of algebraic equations for (a_0, a_1, a_2, b_1, b_2):

$$6a_2m^2 - \alpha\left(a_2^2 - b_2^2\right) = 0, \tag{3.24}$$

$$2a_1m^2 - 2\alpha\left(a_1a_2 - b_1b_2\right) = 0, \tag{3.25}$$

$$-4a_2\left(m^2+1\right) - \alpha\left(a_1^2 - b_1^2 + b_1^2 + 2a_0a_2\right) - \beta a_2 = 0, \tag{3.26}$$

$$-a_1\left(m^2+1\right) - \alpha\left(2a_0a_1 + 2b_1b_2\right) - \beta a_1 = 0, \tag{3.27}$$

$$6b_2m^2 - 2\alpha a_2b_2 = 0, \tag{3.28}$$

$$2b_1m^2 - 2\alpha\left(a_1b_2 + b_1a_2\right) = 0, \tag{3.29}$$

$$-b_2\left(m^2+4\right) - 2\alpha\left(a_0b_2 + a_1b_1\right) - \beta b_2 = 0, \tag{3.30}$$

$$-b_1 - 2\alpha a_0b_1 - \beta b_1 = 0, \tag{3.31}$$

$$2a_2 - \alpha a_0^2 - \beta a_0 - c_1 - \alpha b_1^2 = 0. \tag{3.32}$$

Solving the bove system with the aid of Mathematica or Maple, we have the following solution:

$$\alpha = \frac{6m^2}{a^2},\ \beta = \frac{4\left(a_2m^2 + a_2 + 3a_0m^2\right)}{a_2},\ a_0 = a_0,\ a_1 = 0,$$
$$a_2 = a_2,\ b_1 = 0,\ b_2 = 0,\ c_1 = \frac{2\left(a_2^2 + 3a_0^2m^2 + 2a_0a_2m^2 + 2a_0a_2\right)}{a_2} \tag{3.33}$$

Sothat the solution of Equation (3.20) will be in the form:

$$W(\xi) = -\frac{\beta}{2\alpha} - \frac{2}{\alpha} - \frac{2m^2}{\alpha} + 6\frac{m^2}{\alpha}sn^2(\xi), \tag{3.34}$$

if m →1, we have the hyperbolic solution:

$$W(\xi) = -\frac{\beta}{2\alpha} - \frac{2}{\alpha} - \frac{6}{\alpha}\left(\frac{1}{3} - \tanh(\xi)\right). \tag{3.35}$$

3.2. Example 2. Nano-Ionic Currents of MTs

The nano ionic currents are elaborated in [27] take the form

$$\frac{l^2}{3}u_{xxx}+\frac{z^{\frac{3}{2}}}{l}(xc_0-2ss_0)uu_t+2u+\frac{zc_0}{l}u_t+\frac{1}{l}\left(Rz^{-1}-G_0Z\right)u=0, \tag{3.36}$$

where $R=0.34\times10^9\ \Omega$ is the resistance of the ER with length, $l=8\times10^{-9}$ m, $c_0=1.8\times10^{-15}$ F is the maximal capacitance of the ER, $G_0=1.1\times10^{-13}$ si is conductance of pertaining NPs and $z=5.56\times10^{10}\ \Omega$ is the characteristic impedance of our system parameters δ and x describe nonlinearity of ER capacitor and conductance of NPs in ER, respectively. In order to solve Equation (3.36) we use the travelling wave transformations $u(x,t)=u(\xi),\xi=\frac{1}{l}x-\frac{c}{\tau}t$ with $\tau=Rc_0=0.6\times10^{-6}$ s, to reduce Equation (3.36) to the following non-linear ordinary differential equation:

$$\frac{1}{3}u'''-\frac{cz^{\frac{3}{2}}}{\tau}(xc_0-2sc_0)uu'+\left(2-\frac{zcc_0}{\tau}\right)u'+\left(Rz^{-1}-G_0Z\right)u=0, \tag{3.37}$$

which can be written in the form

$$\frac{1}{3}u'''+H_1uu'+H_2u'+H_3u=0, \tag{3.38}$$

$$H_1=\frac{c}{\tau}B,\ H_2=\left(2-\frac{ce}{\tau}\right),\ B=-z^{\frac{3}{2}}\left(xc_0-2ss_0\right),$$
$$E=c_{0z},\ D=H_3. \tag{3.39}$$

Thus Equation (3.38) takes the form

$$\frac{1}{3}u'''+\frac{c}{\tau}Buu'+\left(2-\frac{ce}{\tau}\right)u'+Du=0. \tag{3.40}$$

Balancing u''' and uu' yields, $N+3=N+N+1 \rightarrow N=2$. Consequently, we get

$$u = a_0 + a_1 sn + b_1 cn + a_2 sn^2 + b_2 sncn, \tag{3.41}$$

where a_0, a_1, a_2, b_1, b_2 are arbitrary constants such that $a_2 \neq 0$ or $b_2 \neq 0$. From Equation (3.41), it is easy to see that

$$u' = dna_1 cn - dnb_1 cn + 2dna_2 sn\ cn - 2dnb_2 sn^2 + b_2 dn, \tag{3.42}$$

$$\begin{aligned} u'' = & -a_1 m^2 sn + 2a_1 sn^3 m^2 + 2m^2 sn^2 b_1 cn - 4a_2 m^2 sn^2 + 6a_2 sn^4 m^2 \\ & + 6m^2 sn^3 cnb_2 - m^2 sncnb_2 - a_1 sn - b_1 cn + 2a_2 - 4a_2 sn^2 - 4b_2 sncn. \end{aligned} \tag{3.43}$$

Substituting Equations (3.41)-(3.43) into Equation (3.40) and equating the coefficients of $sn^4 dn$, $sn^3 cndn$, $sn^3 dn$, $sn^2 cndn$, $sn^2 dn$, sncndn, sndn, cndn and dn to zero, we obtain

$$-8b_2 m^2 - 4\frac{cBa_2 b_2}{\tau} = 0, \tag{3.44}$$

$$8a_2 m^2 + \frac{cB\left(2a_2^2 - 2b_2^2\right)}{\tau} = 0, \tag{3.45}$$

$$-2b_1 m^2 - \frac{cB\left(-3a_1 b_2 - 3a_2 b_1\right)}{\tau} = 0, \tag{3.46}$$

$$2a_1 m^2 + \frac{cB\left(3a_1 a_2 - 3b_1 b_2\right)}{\tau} = 0, \tag{3.47}$$

$$\frac{4}{3}b_2\left(5m^2+2\right) + \frac{cB}{\tau}\left(-2a_0 b_2 - 2a_1 b_1 + 3a_2 b_2\right) + \left(2 - \frac{cE}{\tau}\right)\left(-2b_2\right) + Da_2 = 0, \tag{3.48}$$

$$\frac{-8}{3}a_2\left(m^2+1\right) + \frac{cB}{\tau}\left(2a_0 a_2 - a_1^2 - b_1^2 + b_2^2\right) + 2\left(2 - \frac{cE}{\tau}\right)a_2 + Db_2 = 0, \tag{3.49}$$

$$\frac{1}{3}b_1\left(4m^2+1\right)+\frac{cB}{\tau}\left(-a_0b_1+2a_1b_2+2a_2b_1\right)+\left(2-\frac{cE}{\tau}\right)b_1+Da_1=0, \tag{3.50}$$

$$\frac{-1}{3}a_1\left(m^2+1\right)+\frac{cB}{\tau}\left(2a_0b_1+b_1b_2\right)+\left(2-\frac{cE}{\tau}\right)a_1+Db_1=0, \tag{3.51}$$

$$\frac{-1}{3}b_2\left(m^2+4\right)+\frac{cB}{\tau}\left(2a_0b_2+b_1a_1\right)+\left(2-\frac{cE}{\tau}\right)b_2=0. \tag{3.52}$$

Solving the above system with the aid of Mathematica or Maple, we have the following solution:

Case 1.

$$D=0,\ m=m,\ a_0=\frac{1}{3}\frac{4m^2\tau-2\tau+3cE}{cB},\ a_1=0,\ a_2=\frac{-4m^2\tau}{cB},\ b_1=0,\ b_2=0$$

Case 2.

$$D=0,\ m=\sqrt{\frac{1}{2}-\frac{3cE}{4\tau}},\ a_0=0,\ a_1=0,\ a_2=\frac{-2\tau+3cE}{cB},\ b_1=0,\ b_2=0$$

So that the solution of Equation (3.40) will be in the form:

Case 1.

$$u(\xi)=\frac{1}{3}\frac{4m^2\tau-2\tau+3cE}{cB}+\frac{-4m^2\tau}{cB}sn^2 \tag{3.53}$$

Case 2.

$$u(\xi)=\frac{-2\tau+3cE}{cB}sn^2 \tag{3.54}$$

If m → 1, we have the hyperbolic solution:

Case 1.

$$u(\xi)=\frac{1}{3}\frac{2\tau+3cE}{cB}+\frac{-4\tau}{cB}sn^2 \tag{3.55}$$

Case 2.

$$u(\xi)=\frac{-2\tau+3cE}{cB}sn^2 \tag{3.56}$$

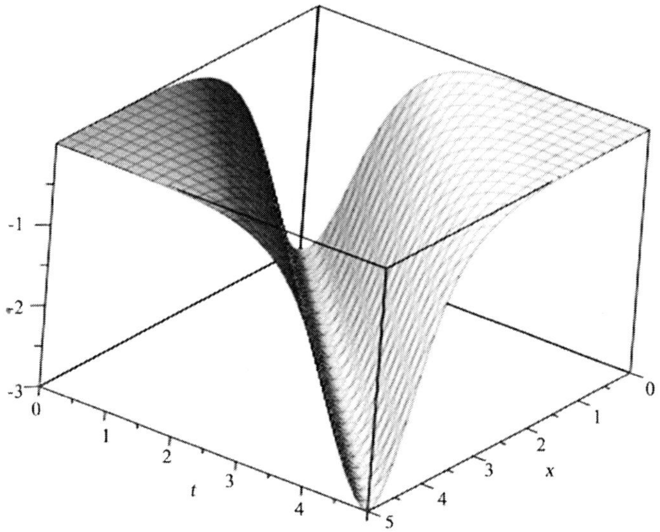

Figure 1. Plot of solution of Equation (3.35).

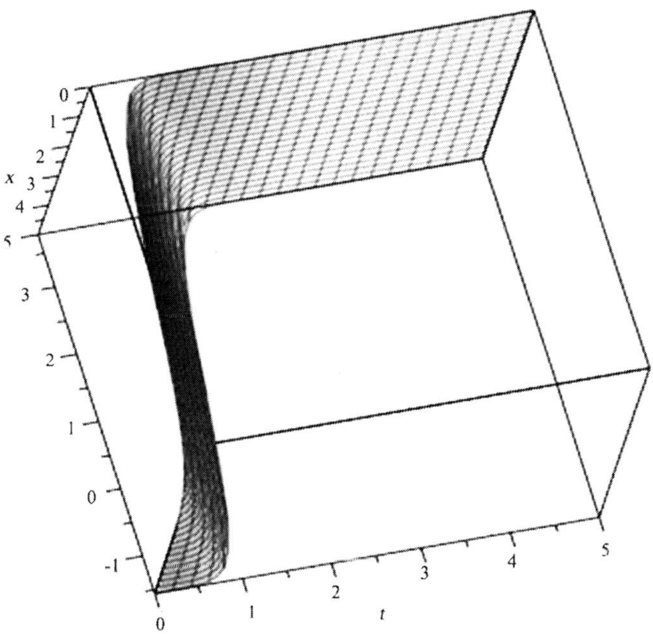

Figure 2. Plot of solution of Equation (3.56).

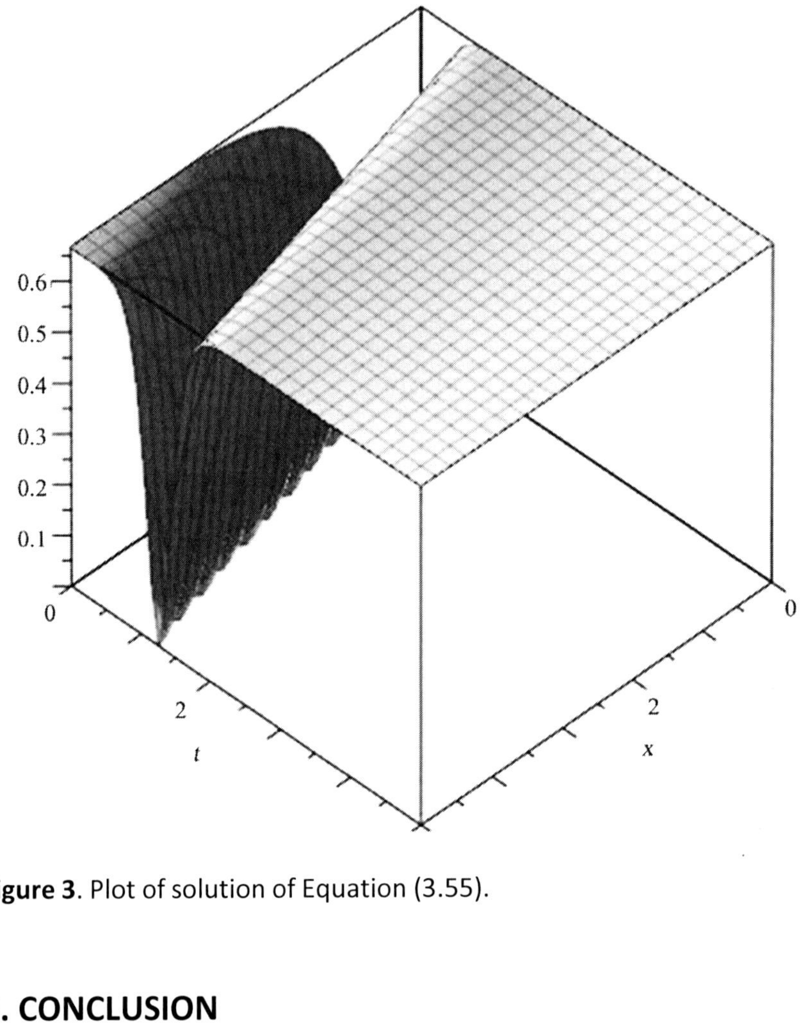

Figure 3. Plot of solution of Equation (3.55).

4. CONCLUSION

The nano waves propagating along microtubules in living cells play an important role in nano biosciences and cellular signaling where the propagation along microtubules shaped as nanotubes is essential for cell motility, cell division, intracellular trafficking and information processing within neuronal processes. Ionic waves propagating along microtubules in living cells have been also implicated in higher neuronal functions, including memory and the emergence of consciousness and we presented an in viscid, incompressible and non-rotating fluid of constant depth (h). The extended Jacobian elliptic function expansion method has been successfully used to find the exact traveling wave solutions of some nonlinear evolution equations. According to the suggested method we obtained a new and more accurate traveling wave solution of nano ionic-solitons waves' propagation along microtubules in living cells and nano-ionic currents of MTs. Let us compare between our results obtained in the present article with the well-known results obtained by other authors using different methods as follows: Our results of nano-solitons of ionic waves propagation along microtubules in living cells and nano-ionic currents of MTs are new and different from those obtained in [27]. Figures 1-3 show solitary wave solution. It can be concluded that this method is reliable and proposes a variety of exact solutions NPDEs. The performance of this method is effective and can be applied to many other nonlinear evolution equations.

CITE THIS PAPER

Emad H. M.Zahran, (2015) Exact Traveling Wave Solutions for Nano-Solitons of Ionic Waves Propagation along Microtubules in Living Cells and Nano-Ionic Currents of MTs. *World Journal of Nano Science and Engineering*,**05**,78-87.

REFERENCES

1. Ablowitz, M.J. and Segur, H. (1981) Solitions and Inverse Scattering Transform. SIAM, Philadelphia.
2. Maliet, W. (1992) Solitary Wave Solutions of Nonlinear Wave Equation. American Journal of Physics, 60, 650-654.
3. Maliet, W. and Hereman, W. (1996) The Tanh Method: Exact Solutions of Nonlinear Evolution and Wave Equations. Physica Scripta, 54, 563-568.

4. Wazwaz, A.M. (2004) The Tanh Method for Travelling Wave Solutions of Nonlinear Equations. Applied Mathematics and Computation, 154, 714-723.

5. El-Wakil, S.A. and Abdou, M.A. (2007) New Exact Travelling Wave Solutions Using Modified Extended Tanh-Function Method. Chaos Solitons Fractals, 31, 840-852.

6. Fan, E. (2000) Extended Tanh-Function Method and Its Applications to Nonlinear Equations. Physics Letters A, 277, 212-218.

7. Wazwaz, A.M. (2007) The Extended Tanh Method for Abundant Solitary Wave Solutions of Nonlinear Wave Equations. Applied Mathematics and Computation, 187, 1131-1142.

8. Wazwaz, A.M. (2005) Exact Solutions to the Double Sinh-Gordon Equation by the Tanh Method and a Variable Separated ODE Method. Computers & Mathematics with Applications, 50, 1685-1696.

9. Wazwaz, A.M. (2004) A Sine-Cosine Method for Handling Nonlinear Wave Equations. Mathematical and Computer Modelling, 40, 499-508.

10. Yan, C. (1996) A Simple Transformation for Nonlinear Waves. Physics Letters A, 224, 77-84.

11. Fan, E.G. and Zhang, H.Q. (1998) A Note on the Homogeneous Balance Method. Physics Letters A, 246, 403-406.

12. Wang, M.L. (1996) Exact Solutions for a Compound KdV-Burgers Equation. Physics Letters A, 213, 279-287

13. Abdou, M.A. (2007) The Extended F-Expansion Method and Its Application for a Class of Nonlinear Evolution Equations. Chaos, Solitons & Fractals, 31, 95-104.

14. Ren, Y.J. and Zhang, H.Q. (2006) A Generalized F-Expansion Method to Find Abundant Families of Jacobi Elliptic Function

Solutions of the (2+1)-Dimensional Nizhnik-Novikov-Veselov Equation. Chaos, Solitons & Fractals, 27, 959-979.

15. Zhang, J.L., Wang, M.L., Wang, Y.M. and Fang, Z.D. (2006) The Improved F-Expansion Method and Its Applications. Physics Letters A, 350, 103-109.

16. He, J.H. and Wu, X.H. (2006) Exp-Function Method for Nonlinear Wave Equations. Chaos, Solitons & Fractals, 30, 700-708.

17. Aminikhad, H., Moosaei, H. and Hajipour, M. (2009) Exact Solutions for Nonlinear Partial Differential Equations via Exp-Function Method. Numerical Methods for Partial Differential Equations, 26, 1427-1433.

18. Zhang, Z.Y. (2008) New Exact Traveling Wave Solutions for the Nonlinear Klein-Gordon Equation. Turkish Journal of Physics, 32, 235-240.

19. Wang, M.L., Zhang, J.L. and Li, X.Z. (2008) The (G'/G)-Expansion Method and Travelling Wave Solutions of Nonlinear Evolutions Equations in Mathematical Physics. Physics Letters A, 372, 417-423.

20. Zhang, S., Tong, J.L. and Wang, W. (2008) A Generalized (G'/G)-Expansion Method for the mKdv Equation with Variable Coefficients. Physics Letters A, 372, 2254-2257.

21. Zayed, E.M.E. and Gepreel, K.A. (2009) The (G'/G)-Expansion Method for Finding Traveling Wave Solutions of Nonlinear Partial Differential Equations in Mathematical Physics. Journal of Mathematical Physics, 50, Article ID: 013502.

22. Zayed, E.M.E. (2009) The (G'/G)-Expansion Method and Its Applications to Some Nonlinear Evolution Equations in Mathematical Physics. Journal of Applied Mathematics and Computing, 30, 89-103

23. Dai, C.Q. and Zhang, J.F. (2006) Jacobian Elliptic Function Method for Nonlinear Differential-Difference Equations. Chaos, Solitons & Fractals, 27, 1042-1049.

24. Fan, E. and Zhang, J. (2002) Applications of the Jacobi Elliptic Function Method to Special-Type Nonlinear Equations. Physics Letters A, 305, 383-392.

25. Liu, S., Fu, Z., Liu, S. and Zhao, Q. (2001) Jacobi Elliptic Function Expansion Method and Periodic Wave Solutions of Nonlinear Wave Equations. Physics Letters A, 289, 69-74.

26. Zhao, X.Q., Zhi, H.Y. and Zhang, H.Q. (2006) Improved Jacobi-Function Method with Symbolic Computation to Construct New Double-Periodic Solutions for the Generalized Ito System. Chaos, Solitons & Fractals, 28, 112-126.

27. Sataric, M., Dragic, M. and Sejulic, D. (2011) From Giant Ocean Solitons to Cellular Ionic Nano-Solitons. Romanian Reports in Physics, 63, 624-640.

Index